Berichte aus dem Institut für Mehrphasenströmungen

Band 9

Fundamentals of microscale bubbles in process engineering

Vom Promotionsausschuss der
Technischen Universität Hamburg
zur Erlangung des akademischen Grades
Doktor-Ingenieur (Dr.-Ing.)

genehmigte Dissertation

von
Simon Matthes

aus
Rüsselsheim

2021

Bibliografische Information der Deutschen Nationalbibliothek
Die Deutsche Nationalbibliothek verzeichnet diese Publikation in der Deutschen Nationalbibliografie; detaillierte bibliographische Daten sind im Internet über http://dnb.d-nb.de abrufbar.

1. Aufl. - Göttingen: Cuvillier, 2021
Zugl.: (TU) Hamburg, Univ., Diss., 2021

1. Gutachter: Prof. Dr.-Ing. Michael Schlüter
2. Gutachter: Prof. Dr. rer. nat. Andreas Liese
Vorsitzender des Prüfungsausschusses: Prof. Dr.-Ing. Dr. h.c. Stefan Heinrich

Tag der mündlichen Prüfung: 17.09.2021

Nonnenstieg 8, 37075 Göttingen
Telefon: 0551-54724-0
Telefax: 0551-54724-21
www.cuvillier.de

1. Auflage, 2021
Gedruckt auf umweltfreundlichem, säurefreiem Papier aus nachhaltiger Forstwirtschaft.

ISBN 978-3-7369-7511-8
eISBN 978-3-7369-6511-9

Vorwort

Die vorliegende Arbeit entstand während meiner vierjährigen Tätigkeit als wissenschaftlicher Mitarbeiter am Institut für Mehrphasenströmungen (IMS) der Technischen Universität Hamburg (TUHH). Die Grundlagen dieser Arbeit entstanden innerhalb des DFG Projektes „Fine Bubbles for Biocatalytic Processes" in Kooperation mit dem Institut für Technische Biokatalyse (ITB, TUHH) und dem Department of Applied Chemistry (Keio University, Japan).

Ich danke meinem Doktorvater Herrn Prof. Dr.-Ing. Michael Schlüter dafür, dass er es mir ermöglichte Teil dieses spannenden, interdisziplinaren Forschungsvorhabens zwischen Deutschland und Japan zu sein. Das mir entgegengebrachte Vertrauen bei der eigenständigen Umsetzung des Projektes und der Repräsentation unseres Forschungsthemas auf internationalen Tagungen hat mich sowohl als Wissenschaftler, als auch als Person reifen lassen. Ihre Begeisterungsfähigkeit während unserer zahlreichen fachlichen Diskussionen wird mir genauso in Erinnerung bleiben wie Ihre herzliche Art.

Danken möchte ich auch Herrn Prof. Dr. rer. nat. Andreas Liese für die gute Zusammenarbeit innerhalb unseres Forschungsprojektes und die Bereitschaft, meine Arbeit als Zweitgutachter zu bewerten, als auch Herrn Prof. Dr.-Ing. Dr. h.c. Stefan Heinrich für die Übernahme des Prüfungsvorsitzes.

Ein großer Dank geht auch an meine Projektpartner Daniel Ohde, Benjamin Thomas, Zeynep Percin sowie Shunya Tanaka und Prof. Dr. Koichi Terasaka.

Während meiner wissenschaftlichen Karriere sind mir viele Menschen begegnet ohne die diese Arbeit nie entstanden wäre: Ich danke Herrn Dr. Patrick Seiler dafür, dass er mich ermutigte den wissenschaftlichen Weg einzuschlagen und durchzuhalten. Meine herausragenden Kollegen, die nicht nur mit fachlichem Rat und Tat zur Seite standen, sondern auch für viele tolle Feiern und Reisen gesorgt haben, haben einen ebenso großen Anteil an dieser Arbeit. Go Penguins! Es stimmt mich positiv, dass viele Freundschaften entstanden sind, welche auch in Zukunft bestehen bleiben. Besonders möchte ich hierbei Dr. Sven Kastens, Felix Kexel, Jürgen Fitschen und Dr. Simeon Pesch erwähnen. Eine große Stütze für mich am Institut waren Bernhard Pallaks und Dr. Marko Hoffmann. Ein großes Dankeschön!

Meinen Studierenden und studentischen Hilfskräften danke ich für Ihre Unterstützung: Maximilian Kamp, Metehan Gürel, Riho Sekimizu, Hashim Ishaque, Akram Rachdi, Malini Dasgupta, Daniela Eixenberger, Miriam Weiß, Yusuke Noguchi, Hendrik Doss, Anosha Zia, Yoko Yamamoto und Timo Merbach. Ich hoffe, dass bei der ganzen Arbeit der Spaß nicht zu kurz gekommen ist.

Zum Abschluss möchte ich mich bei meiner Familie bedanken. Ohne meine Eltern Andrea und Michael, sowie meine Schwester Kathrin wäre ich nicht der Mensch der ich heute bin. Ihr habt mich von Anfang an bei allen Entscheidungen unterstützt und mir alle Möglichkeiten gegeben meinen eigenen Weg einzuschlagen. Danke Annika, dass du mich auch in den schwierigen Phasen ermutigt hast weiter zu machen und mir immer eine Stütze bist.

Hamburg, im September 2021.

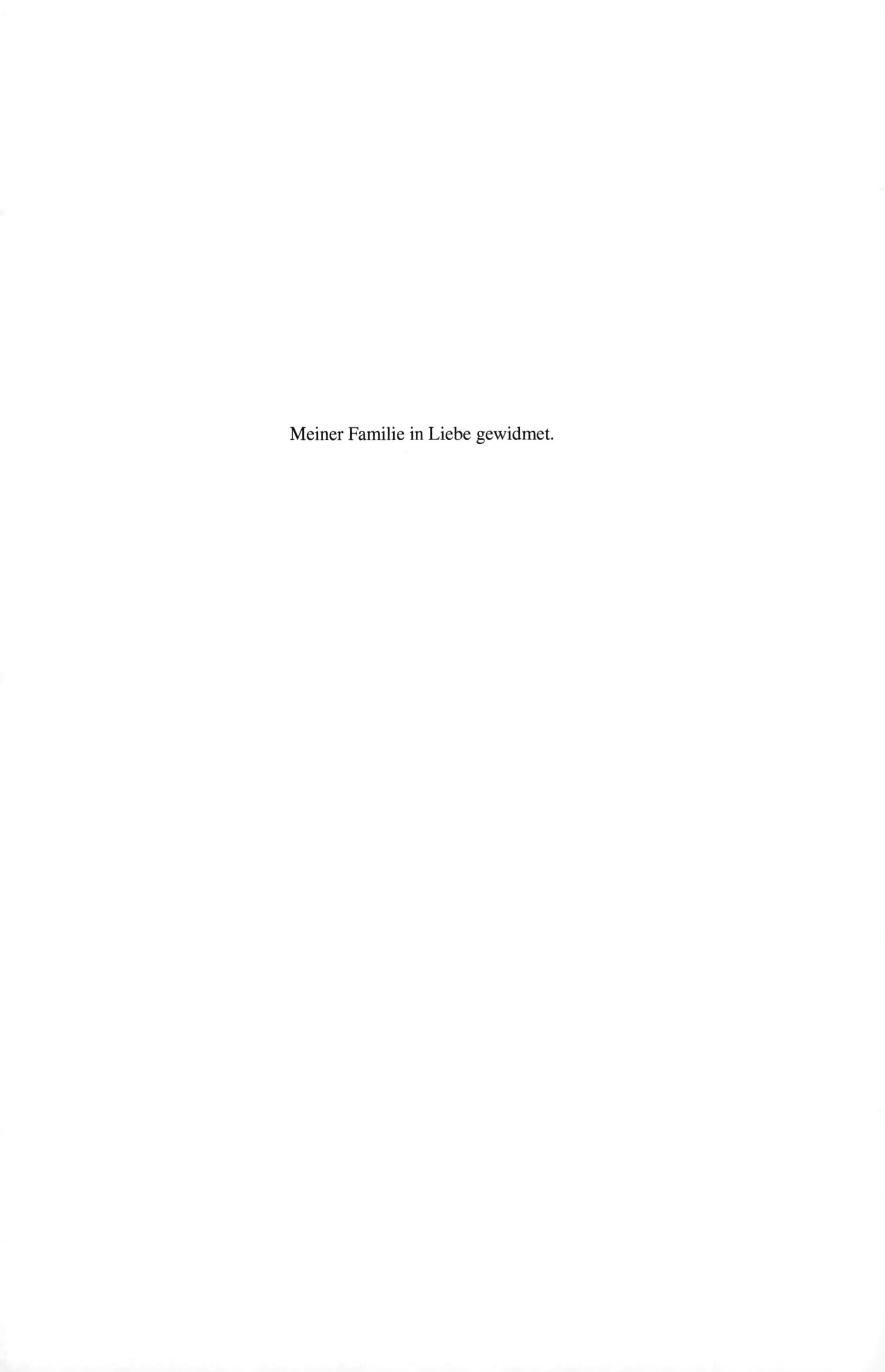

Meiner Familie in Liebe gewidmet.

Table of contents

List of figures

List of tables

Nomenclature

Roman Symbols

A	m^2	surface area
A_0	m	amplitude
a	$m^2 \cdot m^{-3}$	specific interfacial surface area
c	$mol \cdot m^{-3}$	concentration
c_0	mol	initial concentration
c_G	$mol \cdot m^{-3}$	bulk concentration gas phase
c_L	$mol \cdot m^{-3}$	bulk concentration liquid phase
$c_{G,0}$	$mol \cdot m^{-3}$	concentration at gaseous side of interface
$c_{L,0}$	$mol \cdot m^{-3}$	concentration at liquid side of interface
c^*	$mol \cdot m^{-3}$	saturation concentration
D	$m^2 \cdot s^{-1}$	difussion coefficient
D_R	m	reactor diameter
d	m	diameter
d_{32}	m	Sauter mean diameter
d_{50}	m	arithmetic mean diameter
d_B	m	bubble diameter
d_S	m	stirrer diameter

f	Hz	frequency
g	$m \cdot s^{-2}$	gravitational acceleration
H_R	m	reactor height
H^{cc}	–	Henry constant
H^{cp}	$mol \cdot Pa^{-1} \cdot m^{-3}$	Henry constant
h	m	height
h_L	m	liquid level
I	$W \cdot m^{-2}$	intensity
k	$J \cdot K^{-1}$	Boltzmann constant
k_L	$m \cdot s^{-1}$	mass transfer coefficient
M	$N \cdot m$	torque
$N_{pictures}$	–	number of taken pictures
$\dot{N}_i$	$mol \cdot s^{-1}$	mass transfer rate
n	s^{-1}	stirrer frequency
$\dot{n}_i$	$mol \cdot s^{-1} \cdot m^{-2}$	molar flow density
P	W	power
p_{atm}	Pa	atmospheric pressure
p_B	Pa	total pressure inside a bubble
p_{hyd}	Pa	hydrostatic pressure
$p_{partial}$	Pa	partial pressure
p_{YL}	Pa	Laplace pressure
r	m	radial position
S	$kg \cdot m^{-3}$	solubility
s	m	spacing

T	K	temperature
T_1	s	spin-lattice relaxation time
t	s	time
t_c	s	length of ultrasonic pulse cycle
t_p	s	length of ultrasonic pulse
V	m^3	volume
V_{SOPAT}	m^3	measurement volume of SOPAT probe
$\dot{V}$	$m^3 \cdot s^{-1}$	volume flow rate
v	$m \cdot s^{-1}$	velocity
w_g^0	$m \cdot s^{-1}$	superficial gas velocity
x	m	distance
y	m	distance

Dimensionless numbers

$Fo = \frac{4tD}{d_B^2}$	Fourier number
$Ne = \frac{Pn^{-3}}{\rho_L d_S^5}$	Newton number
$Re = \frac{v_B d_B}{\nu_L}$	bubble Reynolds number
$Re_S = \frac{n d_S^2}{\nu_L}$	stirrer Reynolds number
$Sc = \frac{\nu_L}{D}$	Schmidt number
$Sh = \frac{k_L d_B}{D}$	Sherwood number

Greek Symbols

β	$m \cdot s^{-1}$	mass transfer coefficient
χ	m	hydrodynamic diameter
χ_{ratio}	–	ratio of undetected to detected bubbles

Δc	$mol \cdot m^{-3}$	concentration gradient
Δp	Pa	pressure drop
δ_c	m	concentration boundary layer thickness
η	$Pa \cdot s$	dynamic viscosity
η_K	m	Kolmogorov length scale
λ	m	wavelength
ν	$m^2 \cdot s^{-1}$	kinematic viscosity
ρ	$kg \cdot m^{-3}$	density
σ	$kg \cdot s^{-2}$	surface tension
ε	$m^2 \cdot s^{-3}$	energy dissipation rate
ε_G	-	gas hold-up
ζ	-	drag coefficient

Subscripts

0	initial value
B	bubble
fill	filling value
G	gaseous phase
in	inner value
L	liquid phase
max	maximum
O	orifice
out	outer value
R	reactor
rep	reproducibility

res	resolution
S	stirrer
tot	total

Acronyms / Abbreviations

AFM	atomic force microscopy
BSA	bovine serum albumin
BSD	bubble size distribution
CLSM	confocal laser scanning microscopy
CMC	critical micelle concentration
CMOS	complementary metal–oxide–semiconductor
DI	deionized
DLS	dynamic light scattering
DO	dissolved oxygen
FBG	fine bubble generator
FBIA	Fine Bubble Industries Association
FBU	Union of Fine Bubble Scientists and Engineers
ISO	International Organization for Standardization
LED	light-emitting diode
LIF	laser induced fluorescence
MD	molecular dynamics
NMR	nuclear magnetic resonance
NTA	Nanoparticle Tracking Analysis
PMT	photomultiplier tube
SPG	Shirasu Porous Glass

STR	stirred tank reactor
TEM	transmission electron microscopy
TOC	total organic carbon
UFB	ultrafine bubble
UPW	ultrapure water

Abstract

Multiphase contact apparatuses are widely used in the chemical and biocatalytic process industry in which a gaseous reactant has to be supplied for various applications. The gas is mostly supplied by aeration of the reaction medium. Mass transfer occurs at the formed gas-liquid interface, its rate being among the most important parameters characterizing the process. For most processes, the mass transfer rate gets reduced by the low solubility of the gaseous phase within the liquid at ambient conditions, leading to a limitation of the overall process efficiency.

A novel aeration technique based on fine bubbles with diameters smaller than 100 μm, enhancing the mass transfer rate is experimentally analyzed in this work. As a result, models are developed describing the mass transfer mechanisms at microscales which differ from those present in conventionally aerated systems. Due to their small spherical shape, fine bubbles provide a more than 10 times larger volume specific interfacial area than conventional bubbles at the millimeter scale. In combination with their low rise velocity and the resulting high residence time, the impact of the low gas solubility is reduced to a minimum. The potential of using fine bubble aeration in the process industry, is investigated experimentally considering three different reactor concepts: bubble column, jet and stirred tank reactor. The focus of the research is on the bubble dynamics and mass transfer mechanisms at microscopic scales and even smaller.

The experiments within the three different fine bubble aerated reactor concepts lead to the conclusion, that stirred tank reactor setups are most suitable for fine bubble aeration. Therefore, their fundamental hydrodynamics and mass transfer mechanisms are analyzed in greater detail. The results point out the minor role of the impeller on dispersing the gaseous phase. The power input is no longer directly influencing the bubble size distribution, leading to a decoupling of the hydrodynamics within the reactor and its mass transfer characteristics. Compared to conventionally aerated stirred tank reactors, the complexity of the fine bubble aerated system is significantly reduced. With regard to the mass transfer efficiency, the fine bubble aerated stirred tank reactor shows its potential reaching mass transfer coefficients two times greater than those of conventionally aerated systems. Due to their small size, for fine bubbles, counterdiffusion effects show a huge impact on the mass transfer rates reducing the concentration of the process gas within the bubble. Models are developed, describing the decreasing process gas concentration within a bubble over time as well as allowing to calculate mass transfer coefficients in fine bubble flows taking into account the presence of other dissolved gases. Using oxygen, which is a commonly used reactant, single microscale bubbles as well as bubble swarms are investigated. Applying established laser optical measurement techniques, dissolved oxygen concentrations are made visible by fluorescence intensity gradients and quantified in the vicinity of a bubble.

Having shown the benefits of using fine bubbles leads to considering even smaller bubble sizes, offering much higher volume specific interfacial areas. By doing so, the field of ultrafine bubbles with diameters smaller than 1 µm is entered. At these scales, challenges arise due to major differences in thermodynamic and physical properties so conventional theories and models are no longer applicable. The resulting contradictions call into questioning the existence of ultrafine bubbles leading to controversial discussions within the scientific community. The experiments within this work point towards the existence of ultrafine bubbles as well as their long-term stability.

Zusammenfassung

In der chemischen und biochemischen Verfahrenstechnik müssen für eine Vielzahl von Prozessen die unterschiedlichsten Gase als Reaktanten für eine Reaktion bereitgestellt werden. Die Zufuhr des gasförmigen Reaktants wird hauptsächlich durch eine Begasung des Reaktionsmediums gewährleistet, wobei verschiedene Typen von Mehrphasenreaktoren ihre Anwendung finden. Der Stoffübergang aus der Gasphase in das flüssige Reaktionsmedium findet über die sich ausgebildete Phasengrenzfläche statt. Dabei stellt die Stofftransportgeschwindigkeit einen zentralen Parameter für die Charakterisierung eines solchen Prozesses dar. Die geringe Löslichkeit der meisten Prozessgase in Flüssigkeiten bei Umgebungsbedingungen führt zu einer reduzierten Stofftransportgeschwindigkeit, wodurch der Stofftransport zu einem limitierenden Faktor bei der Prozessoptimierung wird.

In dieser Arbeit wird eine neuartige Begasungstechnik, basierend auf dem Einsatz von Feinblasen, experimentell untersucht um höhere Stofftransportraten zu erzielen. Infolgedessen werden Modelle entwickelt, welche die Stofftransportmechanismen auf der Mikroskala beschreiben. Dies ist erforderlich, da sich die Stofftransportvorgänge in feinblasenbegasten Systemen aufgrund der kleineren Dimensionen stark von denen in konventionell begasten Systemen unterscheiden. Feinblasen zeichnen sich durch ihre geringe Größe mit Durchmessern kleiner als 100 µm und ihre ideal sphärische Form aus. Daraus resultiert eine sehr große volumenspezifische Phasengrenzfläche, welche mehr als das Zehnfache von der sich bei konventionellen Blasen im Millimeterbereich ausbildenden Grenzfläche beträgt. Zusätzlich zu der vergrößerten Oberfläche für den Stoffaustausch lässt sich mit Feinblasen eine deutlich höhere Gasverweilzeit erzielen, begründet durch ihre niedrigen Aufstiegsgeschwindigkeiten. Somit wird der Einfluss der geringen Gaslöslichkeit auf ein Minimum reduziert. Das Potential, welches der Einsatz von Feinblasenbegasung für die Verfahrenstechnik bietet, wird experimentell anhand der folgenden drei Reaktorkonzepte untersucht: Blasensäule, Strahl- und Rührkesselreaktor. Die Forschung konzentriert sich auf die Hydrodynamik sowie die Stofftransportmechanismen auf der Mikroskala und kleiner.

Die Experimente zeigen, dass Rührkesselreaktoren am stärksten vom Einsatz der Feinblasenbegasung profitieren. Aus diesem Grund werden die grundlegenden hydrodynamischen und stofftransportspezifischen Eigenschaften dieses Systems genauer analysiert. Die Ergebnisse zeigen auf, dass der Rührer bei vorhandener Feinblasenbegasung eine untergeordnete Rolle für die Dispergierung der Gasphase spielt, wodurch der Leistungseintrag keinen direkten Einfluss auf die Blasengrößenverteilung innerhalb des Reaktors hat. Dies führt dazu, dass die in herkömmlich begasten Systemen interagierenden hydrodynamischen Effekte und die Stofftransportvorgänge für das feinblasenbegaste System separat betrachtet werden können. Dadurch wird die Komplexität des Systems deutlich reduziert. Mit Blick auf die Stofftransportraten wird ein weiterer Vorteil der Feinblasenbegasung sichtbar: Im Vergleich zum herkömmlich begasten System werden im feinblasenbegasten System mehr als doppelt so große Stofftransportkoeffizienten erreicht. Aufgrund des geringen Volumens von Feinblasen führen Gegendiffusionseffekte zu einer starken Verringerung der Gaskonzentration innerhalb der Blase und dadurch zu einer Reduktion der Stofftransportraten. Die in dieser Arbeit entwickelten Modelle ermöglichen eine zeitliche Beschreibung der in der Blase abnehmenden Gaskonzentration, sowie die Berechnung der exakten Stofftransportkoeffizienten unter Berücksichtigung verschiedener gelöster Gase in der Flüssigphase. Am Beispiel von Sauerstoff, welcher für die Oxidation häufig benötigt wird, werden sowohl einzelne Mikroblasen als auch Feinblasenschwärme untersucht. Laseroptische Messmethoden ermöglichen durch Gradienten in der Fluoreszenzintensität die Visualisierung und Quantifizierung des in der Flüssigkeit gelösten Sauerstoffs.

Nachdem die Vorteile der Feinblasenbegasung aufgezeigt wurden, führt die Schlussfolgerung der Ergebnisse zur Untersuchung noch kleinerer Blasen, welche nochmals größere volumenspezifische Phasengrenzflächen aufweisen. Diese Blasen mit Durchmessern kleiner als 1 µm werden als Ultrafeinblasen bezeichnet. In diesem Größenbereich ergeben sich Herausforderungen durch große Unterschiede in den thermodynamischen und physikalischen Eigenschaften der Blasen, sodass herkömmliche Theorien und Modelle nicht mehr anwendbar sind. Die sich daraus ergebenden Widersprüche stellen die Existenz von Ultrafeinblasen in Frage und führen zu kontroversen Diskussionen innerhalb der wissenschaftlichen Gemeinschaft. Die im Rahmen dieser Arbeit durchgeführten Experimente stellen einen ersten Schritt in Richtung des Nachweises der Existenz sowie Langzeitstabilität von Ultrafeinblasen dar.

Chapter 1

Introduction

In today's process industry the efficient supply of oxygen as an oxidant is a central part for various kinds of applications. To utilize oxygen for a process the gas has to be dissolved in the liquid reaction medium. This is mostly done by aeration of the medium where the oxygen is provided by mass transfer across gas-liquid interfaces in multiphase contact apparatuses. The optimization of these complex two-phase reactor systems is still of high interest in terms of saving resources as well as enhancing yields. This study is motivated by the fields of biocatalytic and chemical process engineering. The low solubility of oxygen in liquids under ambient conditions is often limiting the effectiveness of the reaction. For fast chemical reactions in multiphase flows, the mass transfer limitation is usually the bottleneck for a process optimization and becomes crucial with regard to process design and scale up. This results in a high demand for efficient aeration techniques in process industry.

Concerning enzymatic oxidations, not only optimizing the mass transfer is of interest. Most enzymes are sensitive to shear forces which can lead to enzyme denaturation and therefore to a reduced productivity. Commonly used aeration techniques generate macroscopic bubbles with diameters in the order of several millimeters. Here, critical aspects in terms of occuring shear stresses are the bubble formation, the relative velocity between rising bubbles and the liquid phase as well as bubble break-up and coalescence. Also, excessive foaming resulting in a reduced enzyme activity has to be prevented. Therefore, bubble dynamics characterized by the bubble size distribution as well as rising velocities are crucial parameters with a direct influence on the performance of the reactor and therefore a central point for research.

Within this study, a novel aeration technique using fine bubbles is analyzed. The aeration with microscopic bubbles smaller than 100 μm, which can dissolve into the liquid very quickly, is a promising alternative to conventional aeration. Due to their small size and the acting surface tension, fine bubbles appear nearly spherical, resulting in the highest volume specific interfacial surface areas and therefore high mass transfer rates. Furthermore, fine bubbles are characterized by their low rising velocities of just a few millimeters per second due to the small acting buoyancy forces. This leads to smaller induced shear forces at the bubble interface as well as long residence times in the system. The number of bubbles reaching the surface of the liquid phase is reduced compared to macroscopic bubble aeration which benefits the efficiency of the mass transfer and reduces the foaming at the same time.

The potential of using fine bubble aeration in process engineering is investigated experimentally in this work. The results are presented from smallest to largest bubble sizes and with rising level of complexity. The smallest observed bubbles at nanometer scale (ultrafine bubbles) offer the highest volume specific interfaces. In the last ten years, research on ultrafine bubbles has been increased especially in Asia. This relatively young field of research is controversially discussed due to conflicts between reported stabilities of ultrafine bubbles and conventional thermodynamic approaches. A first proof of existence of ultrafine bubbles is adressed in one part of this work as well as an evaluation of their utilization possibility for process engineering.

For determining the efficiency of different fine bubble aeration techniques, different reactor concepts are investigated at laboratory scale with a focus on hydrodynamic and mass transfer characteristics. Therefore, three frequently used reactor systems are considered:

- bubble column reactor
- jet reactor
- stirred tank reactor

For a better understanding of mass transfer effects at microscopic scale, the dissolution process of single fine bubbles is analyzed by means of laser induced fluorescence.

Chapter 2

State of the art and basic knowledge

Within this chapter, the fundamentals on fine bubbles with a maximum diameter of d_B = 100 µm and their applicability for process industry are given. The focus is on clarifying the terminology of fine and ultrafine bubbles as well as characterizing their behavior with regard to hydrodynamics and mass transfer. Therefore, the relevant basics of bubbly flows are described in general and the differences between macrosopic bubbles and fine bubbles are pointed out. In consideration of process applications, the basic knowledge on aerated stirred tank reactors is explained within this context.

2.1 Definition and fundamentals of fine bubbles

The smallest existing bubbles with diameters of several micrometers and smaller are characterized by physical and thermodynamic properties differing from those of macroscopic bubbles. This chapter clarifies the terminology of fine bubbles and points out different possible applications of fine bubble technology. The fundamentals of ultrafine bubbles are discussed in detail due to a lack of a unified accepted theory about their thermodynamic stability.

2.1.1 Classification of fine bubbles

In order to achieve a uniform definition all over the world for the different bubble size ranges, the technical committee 281 "Fine bubble technology" (ISO/TC 281) has been established in 2013 by the International Organization for Standardization (ISO). The ISO/TC 281 has classified fine bubbles with regard to their size and characteristics within the standard ISO 20480. This work follows the recommendation of ISO 20480, where a **fine bubble** is defined as a gas-filled cavity with a diameter smaller than 100 µm. Fine bubbles are further divided into **microbubbles** with diameters larger than 1 µm and **ultrafine bubbles (UFB)** with diameters smaller than 1 µm (compare figure 2.1). In literature, often the term nanobubble is used instead of ultrafine bubble. Due to ISO/TS 80004-1, specifying a nano-object being smaller than 100 nm, the term nanobubble can indicate missinterpretations and is not used within this work. Microbubbles and ultrafine bubbles differ not only in size but also in their physical properties and application possibilities. Therefore, the fundamentals on microbubbles (see section 2.2.3) and ultrafine bubbles (see section 2.1.3) are considered separately. Due to greater similarities in the characteristics of microbbbles and conventional bubbly flows, both scales (microscopic and macroscopic bubbles) are discussed together in chapter 2.2.

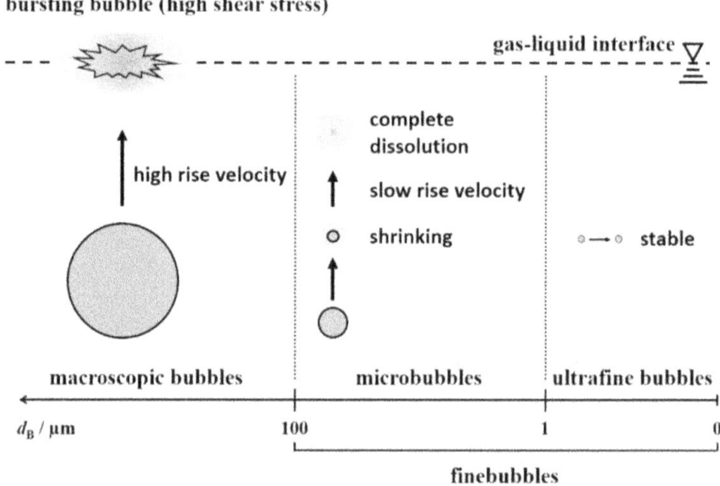

Fig. 2.1 Characteristics and classification of fine bubbles according to ISO 20480 and a comparison with macroscopic bubbles based on [Tho20]

2.1.2 Fields of application for fine bubbles

In the last decades, there has been a strong increase in world wide research on bubbles at diameters of several micrometers and smaller. The fields of application studying these fine bubbles are numerous. In the 1980s and 1990s the potential of fine bubbles as ultrasound contrast agent for medical diagnostics was found [Mil81, DJ91, Sch95, Shi97, Cal98] and later on, microscale bubbles are used for more efficient water treatment by flotation and ozonation [Chu07, Chu08, Bur97, DR94]. Due to their characteristics, fine bubbles have found further applications in recent years. Today, fine bubbles are also used for drag reduction in liquid flows [Ush12, Ush13], cleaning applications [Wu08, Ush11], to improve the fuel efficiency of gasoline engines [Oh13, Nak13] and in aggriculture as well as food industry [Ebi13, Liu13b] or the biochemical process industry [Mah15, Dem12, Doi05, Dru15, Li18, Pet14, Qi02, Yin13, Zim09]. This diversity is reflected in a worldwide increase of research with regard to fine bubbles over the last 40 years. This is quantified by the still growing annual number of scientific papers published on the subject of fine bubbles as shown in figure 2.2.

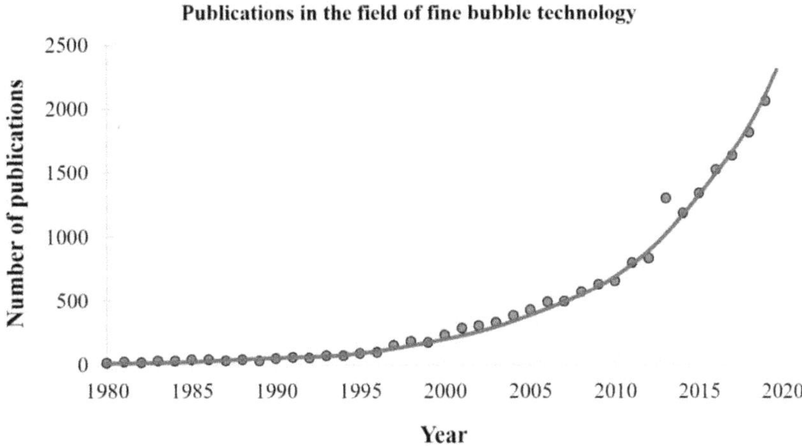

Fig. 2.2 Annual number of papers published on the subject of fine bubbles for the last 40 years

Starting early with the research on fine bubbles, Japan has now a leading role in this field. In Japan, many research groups are working on fine bubbles by focusing not only on fundamental studies, but also on industrial applications. In 2013, the Fine Bubble Industries Association (FBIA), and in 2015, the Union of Fine Bubble Scientists and Engineers (FBU) have been established in Japan to foster the scientific exchange and the collaboration with industrial partners. Initiated by this, more and more reasearch projects on fine bubbles have started outside of Japan during the last years.

2.1.3 Characteristics and physical models describing ultrafine bubbles

The existence of UFB and their application for processes is a controversely discussed topic in todays scientific community [Alh16, Bal12]. The disagreement about the stability of gas bubbles in the nanometer scale is based on the occurance of mass transfer from the gaseous phase inside the bubble into the liquid phase due to a present concentration gradient. Epstein and Plesset [Eps50] as well as Ljunggren and Eriksson [Lju97] have shown, by solving the differential equation governing the shrinkage of a small spherical gas bubble in a liquid phase, that there is no thermodynamic stability of such small bubbles. An UFB inside a liquid describes a metastable system. Any small disturbance is sufficient to disrupt the system, leading to either a shrinking or a growing of the bubble. Even in saturated liquid, a bubble with a diameter of $d_B = 1000$ nm would dissolve in approximately 0.02 seconds [Alh16]. The decisive factor which has to be considered is the surface tension σ which results in an increase of pressure inside the bubble. The total pressure inside a bubble

$$p_{\mathrm{B}} = p_{\mathrm{atm}} + p_{\mathrm{hyd}} + p_{\mathrm{YL}} \tag{2.1}$$

is composed of the atmospheric pressure p_{atm}, the hydrostatic pressure due to the liquid p_{hyd} and the Laplace pressure p_{YL}. According to the Young-Laplace equation, the Laplace pressure

$$p_{\mathrm{YL}} = \frac{4\sigma}{d_{\mathrm{B}}} \tag{2.2}$$

results due to the acting surface tension and increases with a decrease of the bubble diameter. For example, the Laplace pressure of an oxygen bubble in water gets more than ten times the atmospheric pressure at a diameter of $d_B = 70$ nm. Epstein and Plesset [Eps50] have pointed out the influence of the surface tension by showing that only a neglection of the surface tension from the calculation leads to a stable UFB in saturated liquid. For UFB, a distinction must be made between surface UFB, sticking to mostly hydrophobic surfaces, and bulk UFB. As the usable measurements techniques and possible explanation approaches differ for these two kinds, they are considered seperately.

Surface ultrafine bubbles

In contradiction to the theoretical work negating an existence of stable UFB, Parker et al. [Par94a] have found first evidence that nanoscale bubbles are stable for several hours at hydrophobic surfaces. Force measurements between two hydrophobic surfaces moving slowly towards each other in water have shown an increased attraction which cannot be explained by the acting van der Waals force. Additionally, the attraction force is not homogeneously distributed over the surfaces. Steps and discontinuities in the force have been measured along the surface, which is attributed to the presence of nanosclae bubbles at the surface [Att13, Par94a]. By using atomic force microscopy (AFM), a first direct experimental evidence of surface UFB has been found by Lou et al. [Lou00] and Ishida et al. [Ish00] in 2000. The AFM enables a visualization of the gas-liquid-solid three-phase contact line and therefore information about bubble size and contact angle [vL11]. With the use of infrared spectroscopy, Zhang et al. [Zha07] have succeded to identify the substance inside the bubble as well as proving its gaseous state. Besides proving the existence of surface UFB, Zhang et al. [Zha07] could determine the pressure inside the UFB beig in the same magnitude as the atmospheric pressure. There are multiple theories which try to explain the existence of surface UFB by solving the contradiction between the calculations and the measurements. The theories are devided into two groups: one refering to **geometric effects** and the other refering to **thermodynamic effects**.

Geometric effects

A central part of most theories is the correct modeling of the surface tension of an UFB. Zhang et al. [Zha06b] and Attard et al. [Att13] have used the dependency of the contact angle on the surface tension to show that the literature values are not applicable for surface UFB. The contact angles measured by using the AFM data deliver up to 100% larger values compared to the macroscopic advancing contact angle [Zha06b, Zha06a]. This is pointing out a much lower value of the actual surface tension. The geometric effects influencing the surface tension consider the disturbance of the bubble shape and the influence of the curvature [Tol49, Kay86, Fra00]. Since the literature values and most measurement techniques for surface tensions assume a planar interface, there is a source of error. To quantify the change in surface tension due to geometric curvature, Attard et al. [Att13] have used molecular simulations which have shown that this effect is only significant for bubbles smaller than 1 nm. Therefore, **geometric effects** are negligible for the stability of UFB.

Thermodynamic effects

A **thermodynamic effect** is the influence of supersaturation of the liquid on the surface tension [Moo03, Moo04]. As reported by Epstein and Plesset [Eps50], a supersaturated liquid is required for longterm stability of a bubble. Using analytics (virial method) as well as Monte Carlo simulations, He et al. [He05] have pointed out the influence of supersaturation on the surface tension. The surface tension is reduced by a factor of up to five depending on the degree of supersaturation [He05]. Thus, a supersaturation of the liquid has two positive effects on the stability of UFB by reducing the concentration gradient as well as the pressure inside a bubble due to a lower surface tension. As it is hard to achieve an absolute clean liquid, there are also theories which rely on contaminations that adhere to the interface lowering the surface tension [Duc09]. A layer of contamination around the bubble would additionally act as a barrier for diffusion. Another approach relying on thermodynamic effects is the dynamic equilibrium model stated by Brenner and Lohse [Bre08]. For surface UFB, independent of gas type and temperature, the Knudsen number Kn is always larger than one [Sed11, Sed12]. In that case, the fluid dynamics based on a continuum mechanics approach can no longer be used to model the problem. Statistical methods must be used instead. Applying these to a surface UFB, indicate a upward movement of the gas molecules inside the bubble inducing a circulation of the gas-rich liquid close to the surface [Sed11, Sed12]. Benefitted by the flow field, Brenner and Lohse [Bre08] explain the stability of the UFB due to a balance in the incoming and outgoing gas flux of the UFB. This "dynamic equilibrium" ensures that surface UFB are reaching a stable size. The achievement of the dynamic equilibrium is benefitted by the fact that the mass transfer from the bubble into the liquid does not get accelerated by an increasing Laplace pressure. As reported by Lohse and Zhang [Loh15], the curvature of a surface nanobubble reduces during the shrinking process due to the pinning of the contact line, leading to a reduced Laplace pressure.

Bulk ultrafine bubbles

Bulk UFB, being completely surrounded by the liquid phase, differ from surface UFB not only by their ideal spherical shape and the absence of a third solid phase. In contrast to surface UFB, for a bulk UFB the Knudsen number of the gaseous phase is no longer larger than one [Sed12]. Hence, the fluid dynamics and therefore models suitable for surface UFB are not transferable. Though, a stability due to a reduced surface tension caused by either supersaturation of the liquid phase or contaminaton of the interface could be also valid for bulk UFB. One of the biggest challenges in research on bulk UFB is the existing lack of suitable measurement techniques. Regarding the surface tension, there is so far no possibility of a measurement neither directly nor indirectly (as done by using the contact angle dependency for surface UFB). A first approach has been done by Ohgaki et al. [Ohg10] using attenuated total reflection infrared spectroscopy to characterize the hydrogen bonds at the gas-liquid interface of nitrogen UFB in water. Some research on bulk UFB focuses on how the presence of UFB influences liquid properties, which requires highly accurate measurement systems. One approach is the measurement of the liquid density, reducing with rising number density of UFB [Ohg10]. Using nuclear magnetic resonance (NMR) spectroscopy, the paramagnetic effect of water is analyzed. Samples of water including

UFB show an increase in the spin-lattice relaxation time T_1 compared to UFB free samples [Ush10, Liu13b]. The higher T_1 values are attributed to the presence of a gaseous compound changing the physiochemical properties of the sample. A central part of UFB research is finding a suitable measurement technique for their detection and a clear evidence of their gaseous state. Commonly used are laser-based measurement techniques relying on the effect of dynamic light scattering (DLS) [Cho05, Ohg10, Ush10, Wan19]. One of these techniques is the Nanoparticle Tracking Analysis (NTA), delivering information about particle size distributions as well as concentrations (compare chapter 4.1). A disadvantage of these systems is that they cannot distinguish between solid and gaseous phases since both diffract the light similar at the nanoscopic scale. Therefore, further methods are additionally used to classify the physical state of the particles. By instantaneously freezing a droplet containing UFB (using for example liquid nitrogen), freeze-fractured replica are created. Analysis of these replica, using transmission electron microscopy (TEM), identify UFB by their smooth surface in contrast to solid nanoparticles [Ohg10, Uch11]. The freeze-fractured replica also give information about number density distribution and concentration of UFB inside the sample.

Using molecular dynamics (MD) simulations, Weijs et al. [Wei12] have shown that bulk UFB are stable in pure liquid under special conditions. A cluster of UFB stabilizes itself when the bubbles are at a distance smaller than five times their diameter. This shielding effect is produced by the saturation of the liquid in between of the UFB [Wei12]. To have the bubbles stable for long time, bubble collision has to be avoided. One possibility is the addition of salt which slows down the coalescence of UFB at low concentrations [Jin07]. Another possible explanation is the existence of electrostatic repulsion forces between the bubbles to maintain the distance. This explanatory approach is in good accordance to many other theories explaining the stability of bulk UFB due to an electrokinetic potential at the gas-liquid interface. Measurements of the zeta potential are used to quantify the electrical charge of the UFB [Kim00, Cho05]. An accumulation of negatively charged ions at the interface would create a negative pressure. The "coulomb repulsion force" is acting opposite to the surface tension and compensates its effect on the mass transfer [Bun09, Duv12]. Furthermore, Duval et al. [Duv12] make the electric interaction between charged UFB and ions being in solution accountable for the reported long lifetime of UFB up to several days [Ush10, Ebi13, Wei13, Wan19]. One possible explanation for the electrical charge of an UFB is the adsorption of hydroxyl ions at the interface [Kel96, Ush10, Wan19]. In contaminated liquid, Yasui et al. [Yas16] have shown that the dynamic equilibrium model of Brenner and Lohse [Bre08] is also applicable for bulk UFB which are covered with hydrophobic material.

If more than 50% of the bubble surface is covered, the gas enrichment at the hydrophobic material leads to a balance of the incoming and outgoing gas flux.

Especially with regard to bulk UFB many different theories exist, relying on different physical effects. Some models are just applicable under special conditions and as well an interaction of different effects are responsible for the stability of UFB in aqueous solutions. Table 2.1 gives an overview of today's theories explaining an UFB stability and their respective restrictions. The fact of still having no measurement technique for a direct detection of UFB and a missing worldwide accepted theory proving their stability is leading to the existing controversy in the scientific community on that topic.

Table 2.1 Overview of published theories explaining the stability of UFB

physical effect	application	restriction	reference
influence of curvature on σ	surface UFB and bulk UFB	negligible for $d_B > 1$ nm	[Tol49], [Kay86], [Fra00], [Att13]
influence of supersaturation on σ	surface UFB and bulk UFB	maximum reduction of σ up to $1/5$	[Moo03], [Moo04], [He05]
contamination of interface	surface UFB and bulk UFB	no quantification	[Duc09], [Yas16]
dynamic equilibrium of gas in- and outflux	surface UFB	Kn > 1	[Bre08], [Sed11], [Sed12]
diffusive shielding	bulk UFB	theoretical model	[Wei12]
electrically charged surface	bulk UFB	theoretical model	[Kel96], [Ush10], [Duv12], [Wan19]

2.2 Hydrodynamics and mass transfer in bubbly flows

The characterization of multiphase systems is a complex task due to the interaction between the continuous and the dispersed phase. For a precise description and the design of a process information about the hydrodynamics and mass transfer are essential. This chapter focuses on the rising behavior of single bubbles as well as the influence of swarm effects and explains the fundamentals of gas-liquid transport phenomena. In the following, the mass transfer of oxygen from the gaseous into the liquid phase is used as an example.

2.2.1 Fundamentals of rising bubbles in stagnant liquid

In gas-liquid systems, the rise velocity of the bubbles is a crucial parameter due to its direct influence on the residence time of the gaseous phase inside the reactor as well as the total gas hold-up. The residence time of the gaseous phase determines how long the two phases stay in contact and is limiting the amount of transferred gas. For an efficient process, optimal residence times are achievable so that a bubble can contribute most of its oxygen for the reaction. In addition, the rise velocity has an impact on the concentration boundary layer thickness around the bubble which influences the mass transfer behavior (see section 5.4.1).

A free rising bubble is moving with a constant rise velocity (v_B) if the gravitational force, the buoyancy force and the drag force are in equilibrium. The rise velocity is calculated in general as

$$v_B = \sqrt{\frac{4}{3}\frac{|\rho_G - \rho_L|}{\rho_L} g d_B \frac{1}{\zeta}} \tag{2.3}$$

with the gravitational acceleration g, the bubble diameter d_B, the density of the liquid ρ_L and the gaseous phase ρ_G as well as the drag coefficient ζ. The drag coefficient takes the shape and the boundary conditions at the interface of the moving object into account and therefore represents the resistance of the object against its motion. There are large numbers of empirical and semi-empirical correlations available for the drag coefficient [Cli78]. For bubbles the drag coefficient is often expressed as a function of the Reynolds number Re

$$\mathrm{Re} = \frac{\rho_L \cdot v_B \cdot d_B}{\eta_L} \tag{2.4}$$

where the bubble diameter d_B is the characteristic length scale. Figure 2.3 shows how the drag coefficient is changing with increasing Reynolds number for solid particles, droplets and bubbles. For bubbles four regimes are classified corresponding to different bubble shapes and characteristics: Spherical bubbles with an immobile interface (a), spherical bubbles with a moving interface and inner circulation (b), ellipsoidal bubbles with a moving interface and inner circulation (c) and bubbles with a shifting irregular shape (d). Small bubbles are often treated like solid particles with an immobile interface for Re $<< 1$. For those creeping flows, Stoke defines the drag coefficient as $\zeta = 24/\mathrm{Re}$. As long as the bubble stays spherical, the drag coefficient decreases with increasing Reynolds number due to a reduction of the shear stress caused by the inner circulation. If the Reynolds number exceeds a critical value, bubbles are getting deformed due to their mobile interface and the drag coefficient increases. For irregular shaped bubbles, the drag coefficient stays constant as reported by Peebles and Gaber [Pee53].

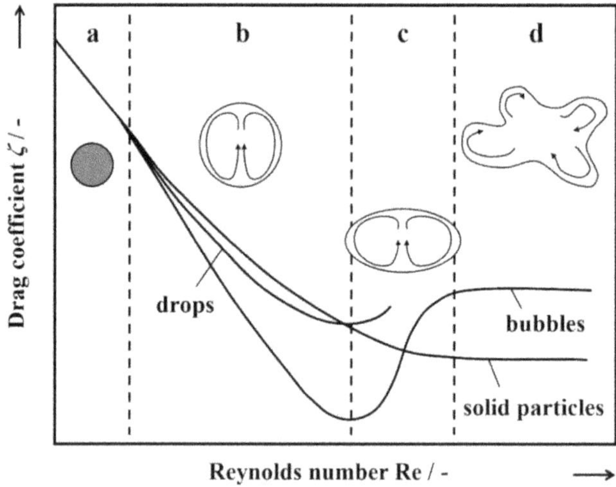

Fig. 2.3 Drag coefficient for bubbles, drops and particles in dependence of the Re number, freely adapted from [Räb13]

For real systems, impurities within the liquid phase influencing the properties of the gas-liquid interface have to be considered. In figure 2.4 this influence is illustrated by comparing the rise velocities for bubbles of different diameters in clean water with those in a contaminated system. For clean water, the rise velocity increases with increasing diameter until the bubble is getting deformed to an ellipsoidal shape, which increases the drag coefficient and lowers the rise velocity. In the contaminated systems, this behavior cannot be seen. The rise velocity increases steadily due to a stabilization of the interface caused by surface active molecules which reduce the mobility of the interface as well as the inner circualtion of the bubble.

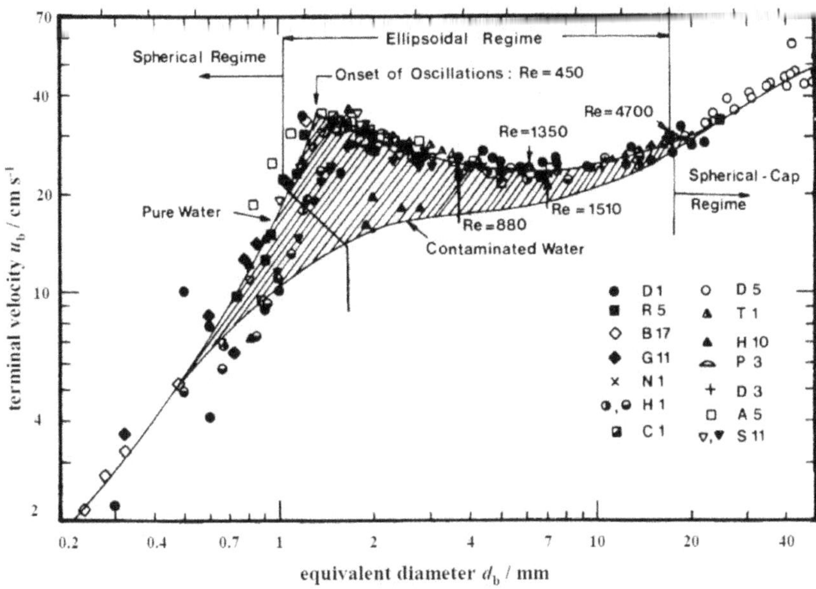

Fig. 2.4 Influence of bubble size and degree of contamination on the bubble rise velocity. Experimental data of air bubbles in pure water, tap water and contaminated water [Cli78]

The so far pointed out results for the rise velocity are valid for a single bubble. Within real processes, dense bubble swarms are present with bubbles of different sizes close to each other, influencing the indidivual rise velocity. In general, the rise velocity in bubble swarms is smaller compared to a single bubble. The reduced cross section of the liquid flow leads to an enhanced momentum transfer, and liquid displaced by the rising bubbles changes the incident flow of the following bubbles [Bra71, Kra12]. Due to the mobile interface of bubbles, the description of their swarm behavior is challenging. Break-up and coalescence mechanisms are changing the bubble size, which also influence the rise velocity. Therefore, the rise velocity of a single bubble is still an important parameter for the characterization of a gas-liquid system. There are many empirical correlations availabe for the rising velocity in particle swarms based on the value for a single particle [Hon84]. The most common equation is provided by Richardson and Zaki [Ric54], developed in 1954.

2.2.2 Mass transfer in two-phase systems

For biocatalytic reactions, where oxygen is transferred into the system by aeration, a quantification of the mass transfer is needed to characterize the process. The three most widely used models describing the mass transfer in gas-liquid systems are based on the film [Lew24], the penetration [Hig35] or the surface renewal theory [Dan51]. Due to its simplicity, the film theory is still applied to bubbly flows and also within this work. As schematically shown in figure 2.5 it is supposed that the total mass transfer takes place in a thin concentration boundary layer with the thickness δ_c and is driven by a concentration gradient Δc. In this laminar boundary layer, mass transfer takes place only by molecular diffusion [Bae94, Kra12]. The gaseous and the liquid bulk phase are well mixed, having a constant concentration.

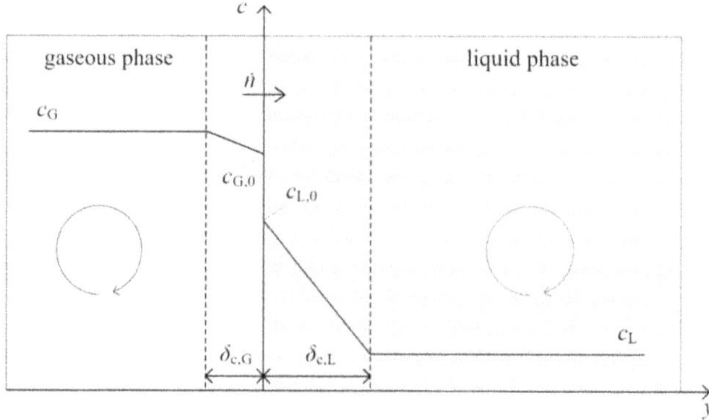

Fig. 2.5 Visualization of the concentration gradients in a gas-liquid system according to the film theory, freely adapted from [Chr09]

For the mass transfer, two physical transport phenomena are distinguished: diffusion and convection. Diffusion is caused by the Brownian motion of molecules. The molar flow density $\dot{n}_i$ of a component i transferred through an interface is given by Fick's law

$$\dot{n}_i = -D\frac{dc}{dy} \tag{2.5}$$

with the diffusion coefficient of component i inside the continuous phase D and its concentration c. The diffusion coefficient influences the speed of diffusion and depends on a variety of factors such as temperature, state of aggregation as well as diffusing species and bulk phase in which the diffusion takes place [Bae94].

With the diffusion coefficient D and the concentration boundary layer thickness δ_c, the mass transfer coefficient β_i

$$\beta_i = \frac{D}{\delta_c} \tag{2.6}$$

is calculated. The mass transfer coefficient is a measure for the mass flow of a component transferred per area and influenced by the flow field, the material properties and the geometric shape of the system as well as the concentration gradient [Bae94]. An analytical solution of equation 2.5 in three dimensions is given by

$$c_i(t,r) = \frac{c_0}{(4\pi Dt)^{3/2}} \cdot exp\left(-\frac{r^2}{4Dt}\right) \tag{2.7}$$

describing the time course of the concentration at a location r when an initial concentration c_0 is applied at $t = 0$ s using a Dirac delta function.

In bubbly flows, usually convective transport phenomena are so strong that diffusion into the bulk phase is negelcted. However, the surrounding fluid is in motion, directly influencing the concentration boundary layer and the concentration within the bulk phase. Hence, convective mass transfer cannot be considered separately due to its interaction with diffusion.

In accordance with the film theory, the molar flow density is calculated for the gaseous and the liquid phase seperately. The molar flow density from the gaseous bulk to the interface is given by

$$\dot{n}_i = \beta_G \cdot (c_G - c_{G,0}) \tag{2.8}$$

with the gaseous bulk concentration c_G and the concentration at the gaseous side of the interface $c_{G,0}$. The molar flow density from the interface to the liquid bulk is defined in the same way to

$$\dot{n}_i = \beta_L \cdot (c_{L,0} - c_L) \tag{2.9}$$

using the concentration at the liquid side of the interface $c_{L,0}$ and the liquid bulk concentration c_L. At the interface, the film theory assumes the equilibrium state described by Henry's law

$$c_{G,0} = H^{cc} \cdot c_{L,0} \tag{2.10}$$

with the Henry coefficient H^{cc} as the proportionality factor.

As shown in figure 2.6, Henry's law is used to calculate the saturation concentration at the liquid side of the interface c_{L}^{*}

$$c_{\mathrm{L}}^{*} = \frac{1}{H^{\mathrm{cc}}} \cdot c_{\mathrm{G}} \tag{2.11}$$

by knowing the corresponding bulk concentration on the gaseous side [Kra12, Sat12, Bra71].

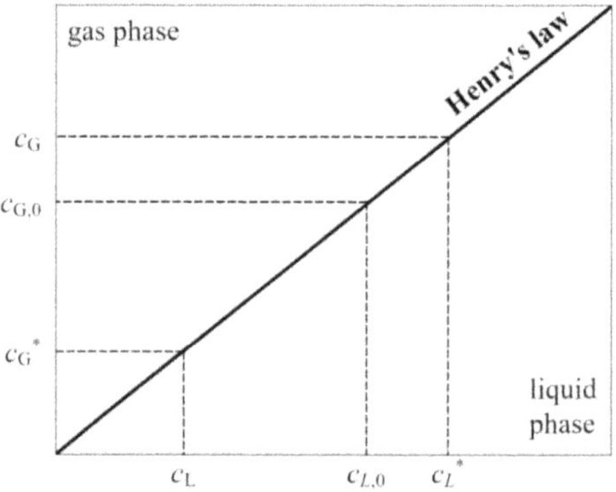

Fig. 2.6 Dependency of the concentration of a component i at the interface of a gas-liquid system according to Henry's law valid for low dissolved gas concentrations in the liquid phase

Using this correlation between the two sides of the interface, the molar flow density is calculated as

$$\dot{n}_{\mathrm{i}} = \left(\frac{1}{\beta_{\mathrm{L}}} + \frac{1}{H^{\mathrm{cc}} \cdot \beta_{\mathrm{G}}}\right)^{-1} \cdot (c_{\mathrm{L}}^{*} - c_{\mathrm{L}}) \tag{2.12}$$

considering both resistances, within the liquid and the gaseous phase, and depending only on the concentration gradient on the liquid side which is easy to measure. The total mass transfer coefficient k_{L} for this case

$$\frac{1}{k_{\mathrm{L}}} = \frac{1}{\beta_{\mathrm{L}}} + \frac{1}{H^{\mathrm{cc}} \cdot \beta_{\mathrm{G}}} \tag{2.13}$$

is definied as the sum of the values for the liquid as well as the gaseous side and is the measure for mass transfer across an interface. For the mass transfer from a dispersed gaseous phase into a continuous liquid phase, the resistance within the bubble is mostly negligible [Sat12, Kra12].

Therefore, $\beta_G >> \beta_L$ so that the total mass transfer rate for a system $\dot{N}_i$ is

$$\dot{N}_i = k_L a \cdot (c_L^* - c_L) \cdot V_R \tag{2.14}$$

with the volume specific interfacial surface area a and the volume of the reactor system V_R. The direct measurement of either the mass transfer coefficient k_L or the volume specific interfacial surface area a is a difficult task and not possible in most systems. That is why the volumetric mass transfer coefficient $k_L a$ is often used for the characterization of the mass transfer performance in mutliphase systems.

For complex processes and the design of industrial reactor systems, the determination of mass transfer performances and influencing flow structures are often not feasible due to a lack of experimental methods, literature data, or from an economic point of view. This leads to a demand of transferring results obtained within simplified systems or from other scales to the desired system. Therefore, the dimensionless Sherwood number

$$\text{Sh} = \frac{k_L \cdot d_B}{D} \tag{2.15}$$

is used to quantify the mass transfer performance for gas-liquid systems [Bra71]. In literature, many empirical correlations for the Sherwood number are found, whose scope of validity is limited by their original use case. Mostly, those correlations are in the form of

$$\text{Sh} = 2 + C_1 \cdot \text{Re}^{\text{a}} \cdot \text{Sc}^{\text{b}} \tag{2.16}$$

where the flow characteristic is expressed by the Reynolds number and the material system by the Schmidt number Sc [Bra71]. The values for the factor C_1 as well as the exponents a and b are determined from experiments. The lower limit for Sh = 2 is given by the analytical solution of a fixed spherical bubble in stagnant liquid [Kra12]. An overview about different Sherwood number correlations for gas-liquid systems is given by Hong and Brauer [Hon84].

In real bubbly flows, the mass transfer is a highly instationary process due to formdynamic bubble surfaces and changing concentration gradients. The decreasing concentration of the transferred species within the dispersed phase during the transport process has been investigated by Streicher and Schugerl [Str77]. They have analyzed the transfer of acetic acid solved in a toluene droplet into the surrounding water. The resulting Sherwood numbers calculated from this experiment are shown in figure 2.7 as a function of the dimensionless time given as the Fourier number $\text{Fo} = 4tD/d_B^2$. The applicability of these results to bubbly flows reaches its limit when the amount of the transferred species is too large to assume a constant bubble diameter over time. Furthermore, a change of the partial pressure that influences the saturation concentration at the interface is not included in this model.

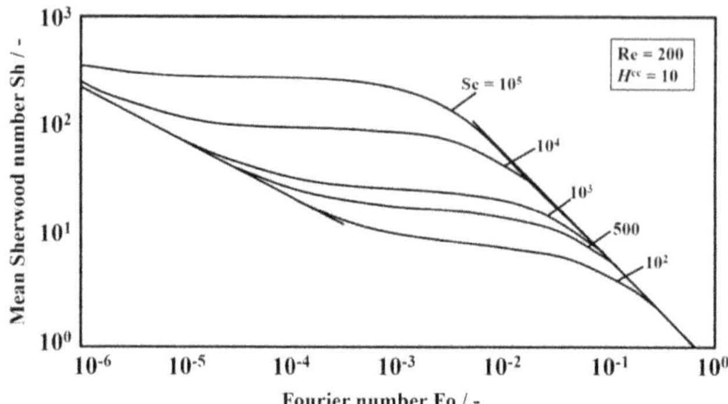

Fig. 2.7 Temporal course of the Sherwood number as a function of the Schmidt and the Fourier number for Re = 200 and $H^{cc} = 10$ freely adapted from [Hon84]

2.2.3 Hydrodynamic and mass transfer characteristics of microbubbles

Microbubbles with diameters smaller than $d_B \leq 100$ µm are the smallest visible bubbles and are described by conventional thermodynamic and physical approaches which have been previously discussed. Due to their small dimensions, microbubbles are characterized by special hydrodynamics as well as mass transfer behaviors distinguishing them from macroscopic bubbles. In contrast to ultrafine bubbles, the aeration with microscopic bubbles is already an alternative technique for mass transfer limited processes. In addition to their large surface to volume ratio, a sufficiently large quantitiy of the gaseous compound is provided for the reaction by achieving gas hold-ups up to 10% [Han17]. In the past twenty years, the potential of utilizing microbubble aeration for the biochemical process industry has been investigated regarding different applications and for various reactor setups.

Concerning biocatalytic and chemical reactions, bubble column reactors, loop reactors, and stirred tank reactors are among the most used reactor systems. For all of these reactor types, first investigations focusing on the performance under microbubble aeration have been conducted so far. The focus of all investigations is the characterization of the dispersed phase as well as an evaluation of the mass transfer in comparison to the conventionally aerated system. Regarding hydrodynamics and mass transfer behavior of free rising microbubble swarms, first investigations in bubble columns have been made by [Bre98, Ter11, Mat18]. Zimmerman et al. [Zim09] and Mahmood et al. [Mah15] are using both experimental approaches and CFD modeling to show the benefit of higher mass transfer rates and better mixing performances for a microbubble aerated airlift loop bioreactor. Concerning stirred tank reactors, the majority of investigations using microbubble aeration are focused on cell cultivation as presented by [Par94a, Qi02, Dru15]. The special hydrodynamic and thermodynamic characteristics of microbubbles that distinguish them from larger ones are explained in the following.

Hydrodynamics of single microbubbles

As discussed for ultrafine bubbles, the Laplace pressure (compare equation 2.2) also plays an important role for microbubbles. Due to the acting surface tension, the inner pressure of bubbles smaller than 100 µm is large enough so that they appear spherical and formstable. Therefore, microbubbles have higher drag coefficients compared to deformable macroscopic bubbles as explained in section 2.2.1. Combined with the small volume of a microbubble, its rise velocity is reduced down to a few millimeters per second, leading to a higher residence time and longer interaction with the liquid phase. Assuming an immobile surface, Stokes' equation [Sto51] for a solid sphere at low Reynolds number (Re << 1) is used for an approximation of the terminal rise velocity of a microbubble v_B

$$v_{B,S} = \frac{d_B^2 \cdot g \cdot (\rho_L - \rho_G)}{18\eta} \tag{2.17}$$

where d_B, g, η, ρ_L and ρ_G are the diameter of the bubble, the gravitational acceleration, the viscosity and the density of the surrounding liquid and the density of the gaseous phase. A further approach for the terminal rise velocity for fluid particles is given by Hadamard [Had11] and Rybczynski [Ryb11], taking into account the reduced capability of a fluid surface to support tangential stress. At a mobile surface, the tangential stress is continuously normal to the interface. The Navier-Stokes equation is solved with boundary conditions considering the internal viscosity of the dispersed phase to define the terminal rise velocity as

$$v_{B,H\text{-}R} = \frac{d_B^2 \cdot g \cdot (\rho_L - \rho_G)}{12\eta} \tag{2.18}$$

which predicts 1.5 times higher values than given by Stokes' equation. Kelsall et al. [Kel96], Parkinson et al. [Par08], as well as Sasic et al. [Sas14] have confirmed the validity of this approach for free rising microbubbles in clean liquids. Only minor impurities inside the liquid phase that get adsorbed at the bubble interface are influencing the rising behavior. The interface becomes more rigid and the rising velocity decreases towards the value given by Stokes' law [Cli78, Sam96, Haa10]. The generation of a completely clean system without any contamination is a time-consuming and expensive process and not realizable for all applications. Therefore the calculation of rising velocities of microbubbles according to

Stoke gives a good assumption for real systems. Not only impurities affect the rise velocity of microbubbles, but also the presence of surfactants within the liquid phase. Surfactants stabilize a microbubble by creating a shell around it, reducing the rise velocity as well [Bre98]. Due to the surfactant, a surface tension gradient is implied that opposes the liquid flow and leads to a no slip boundary condition at the gas-liquid interface [Par11]. Furthermore, the shell formed by the surfactant acts as a mass transfer resistance and simultaneously creates an electric double layer which reduces the coalescence behavior of the bubbles due to electrical repulsion forces [Bre98].

Mass transfer enhancing effect of microbubbles

Also, for the mass transfer behavior of microbubbles, the Laplace pressure is a key parameter which has to be considered in contrast to macroscopic bubbles. Figure 2.8b visualizes the Laplace pressure as a function of the bubble diameter normalized with its value for a 1 mm bubble. It is observed that when entering the fine bubble regime, the Laplace pressure and therefore the total pressure inside the bubble increases exponentially with decreasing bubble diameter. With increasing total pressure inside a bubble, the partial pressure of the gaseous component which is transferred from the bubble into the liquid phase increases as well. As explained in section 2.2.2, the driving force for the mass transfer results from a concentration gradient between the saturation concentration c^* and the concentration in the bulk phase of the transferred component. The saturation concentration is defined as

$$c^* = H^{\text{cp}} \cdot p_{\text{partial}} \tag{2.19}$$

with the Henry constant H^{cp} and the partial pressure of the transferred component p_{partial}. Hence, the saturation concentration is increasing as well during the dissolution process of a microbubble, increasing the concentration gradient and leading to a mass transfer enhancement [Bre98]. Iwakiri et al. [Iwa17] have experimentally studied the mass transfer of single oxygen microbubbles in vacuum-treated water by observing the shrinking process of the bubbles. Figure 2.8a shows that the shrinking rate of a microbubble follows a linear regression until a certain critical size is reached. At a diameter of around 15 µm, the shrinking is accelerated by a factor of four. This change in the shrinking rate implies a 64 times higher mass transfer rate, being explained by the influence of the rising Laplace pressure, which is dominating the mass transfer from that point on. These results are in good agreement with the predicted shrinkage of a microbubble by Worden and Bredwell [Wor98] following an analytical approach using a dynamic model.

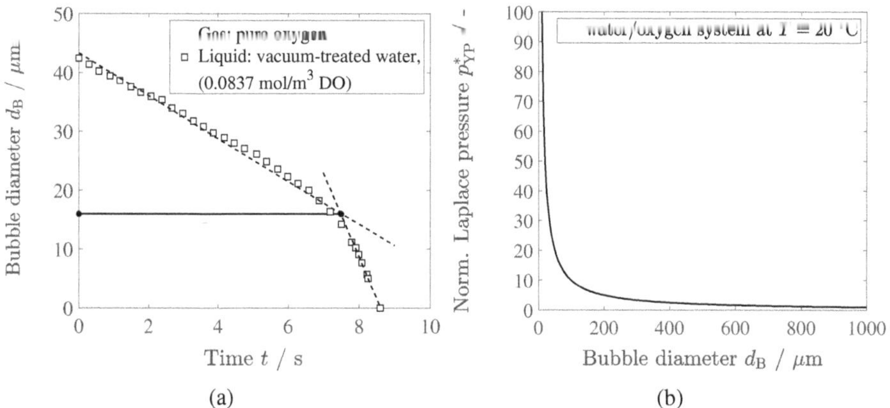

Fig. 2.8 (a) Shrinking rate of a single oxygen microbubble in vacuum-treated water, modified from [Iwa17]. (b) Visualization of the Laplace pressure p_{YL} normalized with its value for a 1 mm bubble $p_{YL}(1\text{ mm})$ as a function of the bubble diameter d_B

Due to this sensitivity of the mass transfer to the bubble size, the pressence of further gaseous components in the system have a larger influence on the mass transfer for microbubbles than for conventionally aerated systems. A mixture of oxygen with a non transferred gas strongly reduces the averaged oxygen mass transfer rate. In figure 2.9 the shrinkage of a microbubble of oxygen and a non transferred gas is shown for different oxygen mole fractions between 20% and 100%. The second gaseous component decreases the shrinking rate and prevents a complete dissolution of the bubble. For example a pure oxygen microbubble has a 14 times higher mass transfer rate transferring 80% of its oxygen compared to an air bubble of the same size transferring the same amount of oxygen. This is important with regard to real processes where other gases are produced as side products during a reaction. Depending on the concentrations in liquid and gaseous phases also a diffusion into the bubble can occur which will affect the mass transfer rate of the microbubbles.

At high bubble densities, the diffusive shielding effect reported by Weijs et al. [Wei13] for ultrafine bubbles (compare section 2.1.3) can also lead to a reduced mass transfer for microbubbles. The presence of neighboring bubbles decreases the concentration gradient for the mass transfer by significantly increasing the background concentration of the bulk phase [Mic18]. This effect has only to be taken into account for stagnant liquids where the shrinkage of the microbubble is faster than the diffusive transport of the dissolved gas within the bulk phase. In multiphase flows, convection is dominating the mass transfer and a diffusive shielding does not occur.

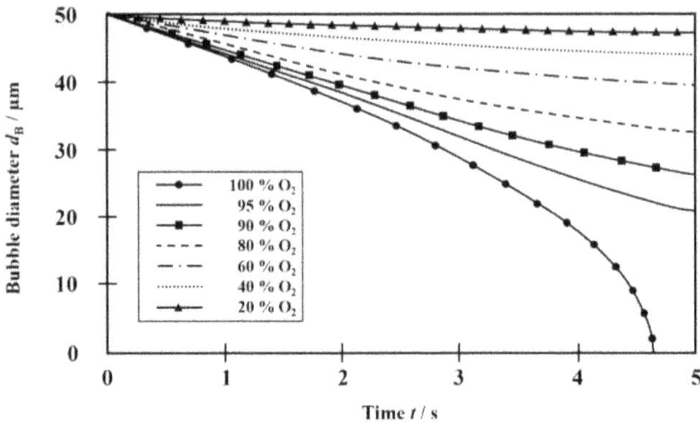

Fig. 2.9 Calculated shrinking rate of a single microbubble of $d_B = 50$ µm filled with an oxygen/nitrogen gas mixture for different oxygen mole fractions [Wor98]

2.3 Suitable reactor systems and generation principles for fine bubble aeration

For an effective application of fine bubble aeration to process industry their generation principle as well as the used reactor system have to be taken into account. Due to its generation principle, each fine bubble generator comes along with advantages and disadvantages making it more suitable for certain reactor systems. As common for standard aeration systems, fine bubble generators as well are classified into dynamic and static aerators [Kra12]. Dynamic fine bubble generators are particularly suitable for using in jet reactors. Jet reactors are characterized by their simple design without any moving parts. The fine bubble nozzle at the top of the reactor is the central part of the system. The mixing of the system is realized by the additional liquid circular flow. In contrast, static fine bubble aerators are used in stirred tank reactors (STR) to replace the classic tube aeration. In the following chapter, the fundamentals on STR, being a central part of this research, and information on the used fine bubble generation principles are given.

2.3.1 Design and hydrodynamics of aerated stirred tank reactors

The aerated stirred tank reactor belongs to the earliest types of reactors in process industry. Even today, STR are most commonly used due to their simplicity, flexibility and low investment costs. About 75% of all reactors for biocatalytic and chemical reactions are still stirred tanks [Kru20]. A STR essentially consist of a cylindrical vessel with optional inlets and outlets for fluid flows, as well as an axially arranged stirrer. A typical key figure for the classification of STR is the ratio of liquid level to vessel diameter h_L/D_R. As they are applied as well for homogenizing, suspending, intensifying, dispersing or emulsifying of the reaction media, the design of STR strongly depends on their primary task [Kra12]. For biocatalytic processes especially high mass transfer rates and a good mixing of the reaction media has to be achieved. Local gradients in temperature and pH value reduce the performance of the enzymatic reaction and influence product quality. Figure 2.10 shows the schematic design of a STR with a single stage stirrer following the standard DIN 28131.

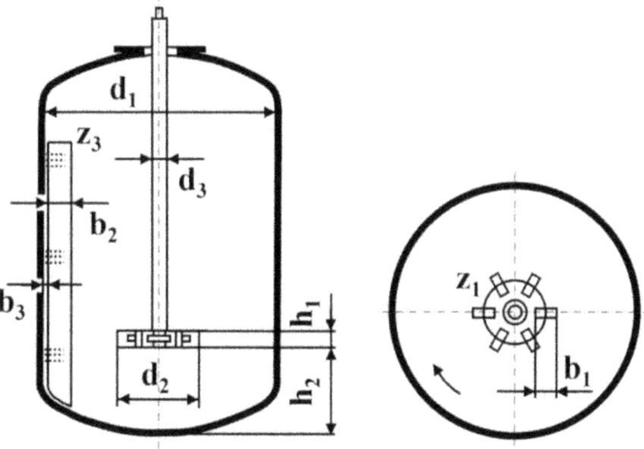

Fig. 2.10 Schematic design of standard STR with a single Rushton turbine following DIN 28131

In addition to the stirrer, baffles are mounted vertically at the reactor wall to modify the flow structure inside the STR. These plates are uniformly distributed over the circumference of the vessel and prevent the liquid from rotating like a solid body as visualized in figure 2.11 by redirecting the liquid flow. In a vessel without baffles, the centrally positioned strirrer causes the liquid phase to rotate as a solid body around the shaft. As a result, a surface vortex is formed. With increasing stirrer speed, the length of the vortex core also increases until it reaches the impeller. At that point, air entrainment occurs resulting in high local shear forces which has to be prevented in shear sensitive enzymatic reactions. Furthermore, the liquid is forced towards the reactor wall due to the solid body rotation and the acting centripetal forces. This reduces enormously the mixing performance of the reactor system. The dimensions of the baffles are defined in the standard DIN 28131 as a function of the vessel diameter. Typically a STR is equipped with three or four baffles. A system with four baffles is defined as "completely baffled" and fully suppresses the solid body rotation of the liquid phase [Kra12, Dor13].

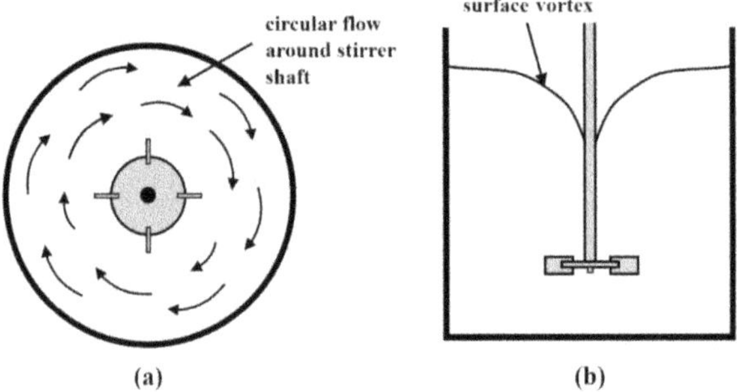

Fig. 2.11 Liquid flow structure in an unbaffled stirred tank. (a) Solid body rotation. (b) Surface vortex [Dor13]

Flow fields and stirrer configurations

In biological processes, often liquids are added to the reaction media at certain times of the process. This is necessary to regulate the pH value during the reaction or to provide nutrients required by the cells. In these fed batch processes the liquid height changes and can lead to a poorly mixed upper compartment and a well-mixed lower compartment in the STR [Ros19]. One major task of the stirrer is to reduce the occurrence of those compartments and to achieve a high degree of homogenization. Due to the variety of processes with different requirements for the mixing, there is a large number of impeller types. Generally, they are classified with regard to the viscosity of the reaction medium and the dominating flow direction induced due to stirring [Zlo01, Kra12, Dor13]. As illustrated in figure 2.12, two main flow directions are distinguished: a radial directed a) and an axial directed flow pattern b). Radially pumping impellers are characterized by a high local fluid velocity at the impeller tip whereby the fluid is radially directed towards the reactor wall. Due to a redirection caused by the wall, an additional axial mixing occurs. In the impeller region, high local shear stress is caused by the high-speed liquid flow. This makes radially pumping impellers as the Rushton turbine beneficial for dispersion of a gaseous phase. A disadvantage of those impellers is that the weak axial flow leads to a reduced mixing time if the liquid level of the STR reaches a certain height.

Axially pumping impellers mainly induce an axial flow pattern inside the vessel leading to good mixing performances. They are favorable for homogenization and suspension of the reaction media. For axially pumping impellers, it has to be distinguished between down-pumping and up-pumping impellers. Down-pumping impellers produce a strong flow towards the bottom of the reactor and thereby extend the residence time of gas bubbles in the system. The longer residence time of bubbles correlates with an increasing mass transfer rate by a better utilization of the gaseous phase [Zlo01, Dor13]. For industrial applications, large $h_{\mathrm{L}}/D_{\mathrm{R}}$ ratios up to $h_{\mathrm{L}}/D_{\mathrm{R}} = 3$ are typical. Therefore, additional impellers are mounted on the stirrer shaft to maintain a good mixing of the whole reactor volume. In those multi-stage systems often different impeller types are used. To enhance the performance of the reactor, the dispersing abilities of radially pumping impellers are combined with the good mixing characteristics of axially pumping impellers [Kru20].

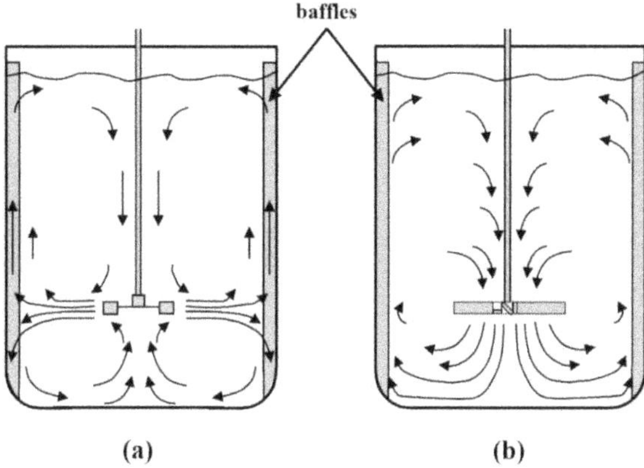

Fig. 2.12 Liquid flow structure in a baffled stirred tank for a radially pumping impeller (a) and an axially pumping impeller (b) [Dor13]

According to Judat [Jud76], standard impeller types for low viscous media are used for liquid viscosities of $\eta_L < 500$ mPa · s. The biocatalytic processes which are considered within this work are assigned to the low viscosity regime, having dynamic vicosities in the same magnitude as water ($\eta_L \approx 1$ mPa · s, depending on temperature). Figure 2.13 displays the three impeller types which are considered for the characterization of fine bubble aerated STR (compare chapter 5.3). The **Rushton turbine** is the most used impeller type in bioreactors. In approximately 90% of all STR this radial pumping impeller is applied for the dispersion of the multiphase system [Kru20]. It consists of a disc which diameter is 1/3 of the reactor diameter ($d_S/D_{\mathrm{R}} = 1/3$). At the edge of the disc, six blades are perpendicularly mounted and equally spaced over its circumference.

Representative for axially pumping impellers, a standard down-pumping **pitched blade** (six blades at 45-degree angle) and a **segment impeller** are investigated as well. The segment impeller is characterized by its low local shear forces and is therefore preferably used for shear sensitive processes as the cultivation of animal or plant cells [Kru20].

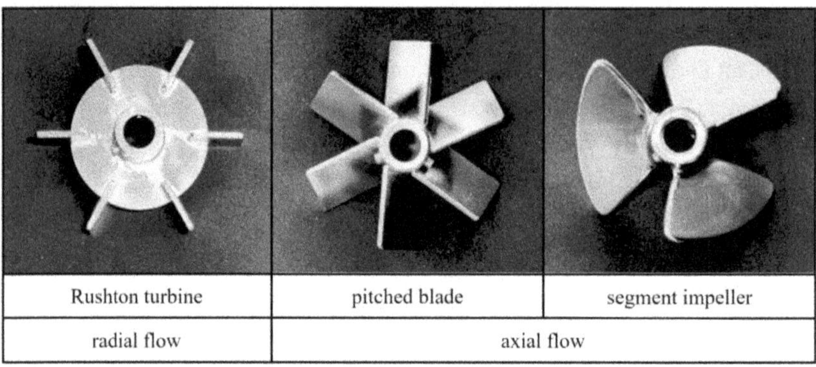

Fig. 2.13 Rushton turbine, pitched blade and segment impeller classified by their induced flow direction

Power input and flow regimes

For the characterization of stirred systems, dimensionless numbers are used to describe the flow regime and the power input as characteristic parameters for the agitated tank. Using dimenesionless numbers, the complexity of a problem is reduced by minimizing the input parameters. At the same time the comparability of different scales and configurations is enabled. The flow regime within a stirred tank is described by the Reynolds number $\mathrm{Re_S}$

$$\mathrm{Re_S} = \frac{n \cdot d_\mathrm{S}^2}{\nu_\mathrm{L}} \tag{2.20}$$

as a function of the stirrer frequency n, the diameter of the stirrer d_S and the kinematic viscosity of the reaction medium ν_L. According to Zlokarnik [Zlo01], for STR the flow is considered as turbulent for Reynolds numbers $\mathrm{Re_S} > 10000$. Due to the low viscosity of the reaction medium, in the range of water, the biocatalytic reactions considered in this work always operate within the turbulent regime, also at low stirrer frequencies.

The power P which is induced to the system by the impeller is a key parameter for characterizing the process and is calculated as

$$P = 2\pi \cdot M \cdot n \quad (2.21)$$

with the stirrer frequency n and the torque at the stirrer shaft M. The measurement of the torque is particularly challenging because the energy loss due to bearings, couplings and inside the gear has to be considered. This is important, calculating the necessary motor power for a specific process. The amount of energy induced to the liquid is the relevant process parameter and is given as dimensionless power input by the Newton number Ne

$$\mathrm{Ne} = \frac{P}{\rho_\mathrm{L} \cdot n^3 \cdot d_\mathrm{S}^5} \quad (2.22)$$

using the density of the liquid ρ_L, the stirrer frequency n, and the impeller diameter d_S. Additionally, the Newton number is influenced by the geometric dimensions of the reactor as the $h_\mathrm{L}/D_\mathrm{R}$ ratio or the installation of baffles as well as the stirrer configuration. Both the impeller type and their combination have to be taken into account. The diagram in figure 2.14 shows the Newton number as function of the stirrer Reynolds number for a single stage stirrer system with a ratio $h_\mathrm{L}/D_\mathrm{R} = 1$. The Newton number is determined for the most commonly used impeller types for a baffled and an unbaffled system. For all investigated impellers, in the laminar flow regime, the Newton number strongly decreases with increasing Reynolds number. Reaching the critical Reynolds number at $\mathrm{Re_S} = 10000$, the Newton number stays constant within the turbulent flow regime. In general two significant effects are identified: Firstly, radially pumping impellers have higher Newton numbers compared to axially pumping impellers. Thus, these so called high torque impellers induce the same power input as an axially pumping impeller at lower stirrer frequencies. Secondly, the installation of baffles leads to higher Newton numbers. For example the Newton number of $\mathrm{Ne} = 0.2$ of an unbaffled impeller system increases to $\mathrm{Ne} = 0.75$ for the completely baffled system [Zlo67, Zlo01, Dor13, Bat63, Buj87].

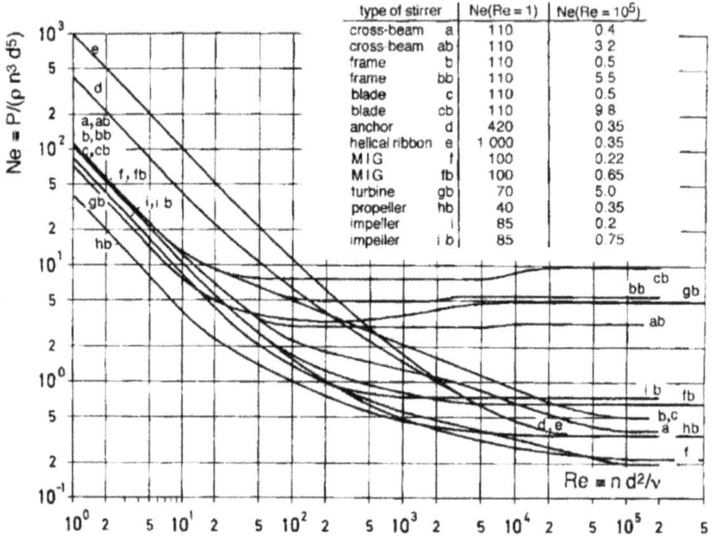

Fig. 2.14 Newton number in a STR as a function of the stirrer Reynolds number determined for various stirrer configurations [Zlo67]

Aeration mechanisms

Many chemical and biocatalytic processes are consuming a gaseous phase to maintain the reaction. In most cases the gaseous phase is applied to the process by bubble aeration. In STR, static aerators are used for the bubble formation which do not need an additional external energy input [Kra12]. The bubbles are primarily formed by the orifices and pores of the aerator through which the gaseous phase enters the system. Figure 2.15 illustrates four commonly in process engineering used static aerators which are applied in bubble columns as well as STR. With an open tube, the gaseous phase is localy supplied to the rection media leading to a strong backmixing of the two phases and a heterogeneous distribution of the bubbles over the reactor cross section. Due to its simple design and good cleaning abilities, the open tube is still often implemented in industrial applications. Ring spargers and perforated plates generate a more homogeneous distribution of the gaseous phase, and sintered spargers can additional reduce the initial bubble size by a multiple. Sintered spargers come along with the disadvantage of fouling, which can occur during the process, and therefore are less popular in large-scale industry [Bot16]. In STR the bubble formation is strongly influenced by the stirrer and the resulting liquid flow inside the vessel. Two mechanisms have to be taken into account: First, the acting shear forces at the sparger caused by the liquid flow favor the detachment of smaller bubbles resulting in a size reduction. The bubble formation is no longer only a function of gravitational and surface forces. Furthermore, the direct interaction between the impeller and the gaseous phase contributes to the dispersion as well as the distribution of the gas inside the reactor. With regard to an efficient reaction, the generation of small, homogeneous distributed bubbles has to be achieved, to realize an intensive mass transfer performance [Kra12].

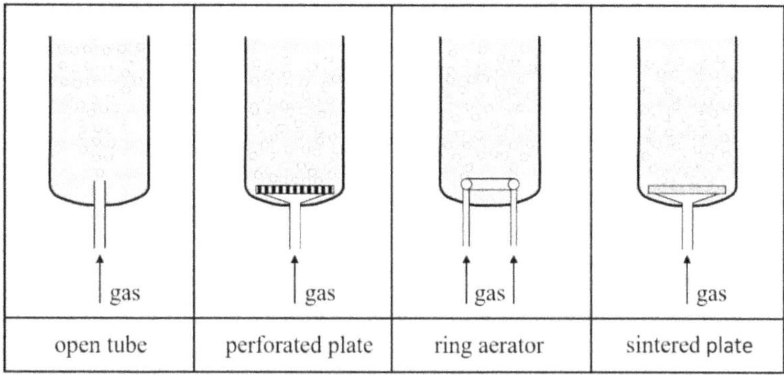

Fig. 2.15 Illustration of four commonly in process engineering used static aerators according to [Kra12]

Gas hold-up and specific interfacial surface area

Another parameter for the characterization of gas-liquid systems is the gas hold-up ε_G. Due to its correlation to the bubble size distribution and the system properties, the gas hold-up is an elementary parameter for the description of the gas dispersion. The gas hold-up

$$\varepsilon_G = \frac{V_G}{V_{tot}} = \frac{V_G}{V_G + V_L} \tag{2.23}$$

is defined as the ratio between the volume of the gaseous phase V_G and the total volume of the two-phase flow V_{tot}. A multitude of design and operating parameters are influencing the gas hold-up, and the experimental determination is still challenging. A good approximation for the gas hold-up is given by the proportionality

$$\varepsilon_G \propto (w_g^0)^n \tag{2.24}$$

to the superficial gas velocity w_g^0. For a homogeneous flow regime, which mostly is the case in fine bubble flows, the exponent n is close to one [Bot16].

Even in a homogeneous flow regime, the assumption of a monodispersed gaseous phase with a constant bubble diameter is an ideal concept. In real systems the bubble sizes vary within a wide range which is visualized using distribution functions. To describe these dispersed systems as simple and precise as possible, different mean diameters are used. With regard to the mass transfer of a two-phase system, both the surface and the volume of the gaseous phase are of importance. To take both values into account, J. Sauter developed in 1926 the Sauter mean diameter

$$d_{32} = 6 \cdot \frac{V_{\mathrm{G,tot}}}{A_{\mathrm{G,tot}}} \tag{2.25}$$

as a function of the total volume of the gaseous phase $V_{\mathrm{G,tot}}$ and the total surface of the bubble swarm $A_{\mathrm{G,tot}}$ [Wan13, Bou01, Alv02]. A monodispersed model system with all bubbles at the size of the Sauter mean diameter equals the real system in the total volume and total surface area of the dispersed phase.

Having the information on the gas hold-up and the Sauter mean diameter, the specific interfacial surface area of the system

$$a = \frac{6 \cdot \varepsilon_{\mathrm{G}}}{d_{32}} \tag{2.26}$$

is defined by their ratio. The specific interfacial surface area is an essential parameter for the evaluation of mass transfer performances. It describes the area through which the mass transfer between the two phases takes place.

2.3.2 Fine bubble generation principles

The formation of fine bubbles is realized by utilizing different physical and fluid dynamical effects. Primarily, the pressure dependency of the gas solubility in liquids or the interaction of the gaseous phase with local shear gradients is used to generate high microbubble densities. Almost all fine bubble generators produce a wide spectrum of bubble sizes reaching from a few nanometers (**ultrafine bubbles**) to several micrometers (**microbubbles**). In contrast to ultrafine bubbles which are stable for a certain period of time, the generated microbubbles leave the liquid system by rising to the surface, or dissolve completely in the liquid.

In general, the different generation principles are classified into the groups of static or dynamic aeration. Most commercially available fine bubble generators belong to the category of dynamic aeration mechanisms. They are characterized by producing high dense fine bubble flows. Therefore, an additional liquid flow has to be afforded for the bubble formation, which places special demands on the setup. In contrast to this, static aerators are easily applied to any system due to their simple design. In the following, the working principles of the different fine bubble generators used within this work are explained in detail.

Porous spargers (static aeration)

As mentioned in chapter 2.3.1, **sintered spargers** are used to achieve a homogeneously distributed gas phase with the smallest bubble diameters. Due to its low power consumption, the fine bubble generation through porous materials is a promising technology for technical applications [Han17, Neh04]. Today's manufacturing processes enable the fabrication of spargers with mean pore sizes smaller than 1 µm. Using these spargers in liquids with low surface tension results in fine bubble aeration with microscopic bubbles of around 100 µm in diameter [Ohd19, Tho20, Mat20]. In combination with shear forces acting on the spargers surface due to a fluid motion, even smaller bubbles are produced. Therefore, this sparger is especially suitable for the fine bubble aeration in STR [Qi02, Dru15, Mat20]. Instead of sintered spargers, in the field of fine bubble technology **membranes** made of Shirasu Porous Glass (SPG) are also used for the generation of microbubbles. Developed in 1981 by the Miyazaki Prefectural Industrial Laboratory, SPG is characterized by uniform pore diameters which can be accurately adjusted between 0.1 µm and 20 µm. Furthermore, SPG is resistive to mold and bacteria, making it favorable for biological applications. Its good formability enables different designs customized for the used system including surface modifications for a better wettability [Kuk06, Kuk09, Liu13a].

Ultrasonic irradiation (static aeration)

In gas-saturated liquids, fine bubbles are generated utilizing the effect of cavitation. Therefore, precise pressure changes have to be induced inside the liquid phase to overcome the existing free-energy barrier and to start the nucleation of small, gas-filled cavities [Jon99, Wan19]. Applying ultrasonic irradiation with frequencies in the kHz range, the bubble nuclei are growing due to the periodic pressure changes inside the liquid. The growth continues until the resonance bubble diameter is reached at the size of a few micrometers [Lei94]. After reaching the resonance diameter, the cavitation bubbles will collapse and fragment into smaller bubbles which start growing again [Sus89, Yas19]. Therefore, the number concentration of fine bubbles increases with irradiation time towards a maximum value. With decreasing frequency and increasing power of the ultrasonication, the maximum number concentration increases as well [Kim00, Cho05, Yas19]. At the same time, higher power values are resulting in larger diameters of the produced UFB [Cho05]. The advantages of ultrasonic irradiation for fine bubble generation is given by a compact and simple design, working without any moving parts, what guarantees a contamination-free operation. Controlling the irradiation frequency and the power input of the sonication, the size of the genrated bubbles is adjusted so that only ultrafine bubbles of a specific size are produced.

Pressurized dissolution method (dynamic aeration)

The pressurized dissolution method is used for the generation of fine bubbles smaller than $d_B = 50$ µm with high number densities of up to 10^6 bubbles/mL [Mae15, Tsu14]. Its working principle is based on the outgassing of saturated liquids due to an abrupt decompression. As visualized in figure 2.16, liquid and gaseous phases are mixed together and pressurized within a tank to increase the concentration of dissolved gas. After the pressure tank, the mixture is pumped through a nozzle into a basin at ambient pressure, where the fine bubbles are formed. Due to the pressure drop, the liquid becomes oversaturated and cavitation bubbles are generated. Bubble sizes and number densities are controlled by the overpressure inside the tank [Ter11]. By circulation of the liquid, fine bubbles are generated continuously. The concentration of bubbles increases over time until a maximum value is reached.

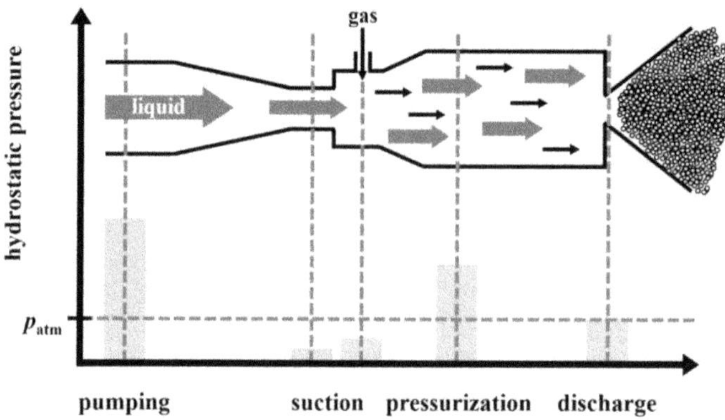

Fig. 2.16 Schematical illustration of the operating principle for the pressurized dissolution method as developed by IDEC, Japan

Swirl type (dynamic aeration)

As shown in figure 2.17, fine bubbles are generated by the application of a spiral liquid flow [Ter11]. Therefore, the liquid phase is pumped tangentially into the cylindrically formed bubble generator. According to Bernoulli's theorem, the circulating flow leads to a negative pressure along the center of the mixing chamber, what causes an automatical suction of the gaseous phase. The gas-liquid two-phase flow leaves the mixing chamber through a small orifice at the top of the bubble generator. The dispersion into microscopic bubbles is caused by centrifugal effects inside the bubble generator and the acting shear forces at the outlet [Ter11, Tsu14]. The achievable bubble sizes and concentrations using this principle are controlled by adjusting the volume flow rate of both the liquid and the gaseous phase. In 1999, Onari et al. [Ona99] first used this fine bubble generation type for the aquaculture of oysters. They have reported a very sharp distrubtion of fine bubbles with a mode value of 15 μm.

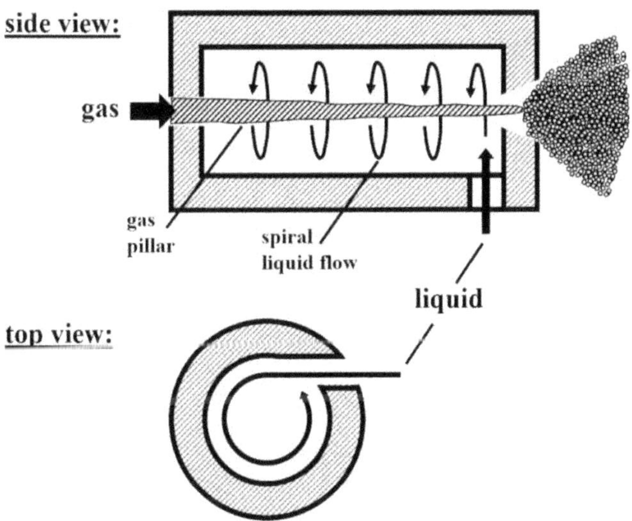

Fig. 2.17 Operating principle of a swirl type fine bubble generator [Ter11]

Ejector type (dynamic aeration)

For ejector type fine bubble generators, the liquid phase creates the driving jet being pumped straight through the generator. As illustrated in figure 2.18, the mixing tube is the central part of the bubble generator. Due to the pressure drop caused by the mixing tube, the gaseous phase is sucked inside the system. The distance of the mixing tube from the exit plane of the nozzle strongly influences the ejector performance [Nag92]. The dispersion of the gaseous phase and further reduction of the bubbles is induced by the turbulent mixing inside the mixing tube and the shear forces within the expansion region of the jet [Nak13]. Compared to the other fine bubble genaration principles described within this chapter, an ejector is generating the largest bubble sizes of a few hundred micrometers [Nag92, Tsu14].

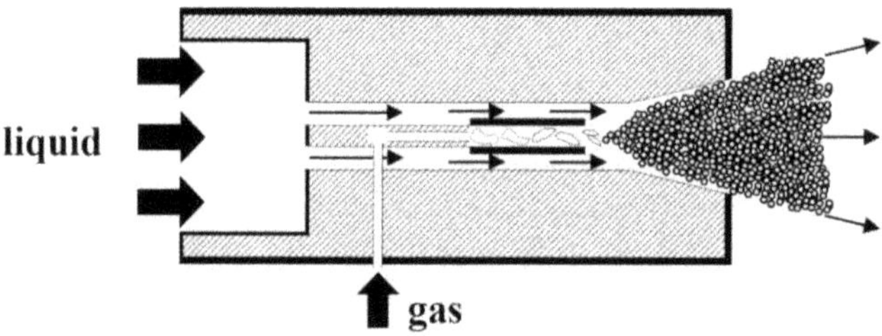

Fig. 2.18 Operating principle of an ejector type fine bubble generator [Ter11]

Chapter 3

Experimental setups and procedure

This work deals with the fundamentals of fine bubbles and their applicability to process industry. The characterization with regard to achievable bubble sizes as well as hydrodynamic and mass transfer effects within a fine bubble two-phase flow are of interest. The experimental approach is classified into three parts:

- proof of stability for UFB and evaluation of their benefit for process engineering
- evaluation of the performance of fine bubble aerated reactor systems
- analysis of fundamental mass transfer processes at microscale bubbles

In the first part, the existence and stability of bulk UFB is critically examined by comparing two different generation mechanisms. Therefore, UFB concentrations and UFB diameters are determined for ultrasonication as well as pressurized dissolution.

Within the second part, the impact of fine bubble aeration on the performance of **bubble column**, **jet** and **stirred tank reactors** is investigated. The bubble dynamics are analyzed for all three systems and a comparison with regard to mass transfer performances is carried out.

For a better understanding of the acting mass transfer mechanisms of fine bubbles, the fundamental mass transfer processes at microscale bubbles are analyzed in detail in the third part. In addition to the investigation of the mass transfer performance of the fine bubble aerated reactor system, a deeper insight into mass transfer phenomena at single microscale bubbles is given, visualizing local dissolved oxygen (DO) concentrations. A challenge of mass transfer analysis at the microscopic scale is that both high temporal and spatial resolutions are required. In section 3.3 the setup for generating single microscale bubbles as well as two different experimental approaches for the measurement of local DO concentrations are explained.

3.1 Proof of stability for ultrafine bubbles in aqueous liquids

As described in chapter 2.1.3, the existence of longterm stable UFB in aqueous solutions is still a controversially discussed topic. To evaluate the stability of UFB based on own results, UFB generated via ultrasonication and pressurized dissolution are analyzed. Both generation techniques are compared with regard to the produced UFB concentration and size. In the following sections, the two setups for the two different UFB generation techniques are explained in detail.

To avoid interferences caused by contaminations, ultrapure water (UPW) is used for all UFB experiments. The UPW is produced using an Evoqua Ultra Clear GP system which guarantees a conductivity of 0.055 μS/cm and a maximum total organic carbon (TOC) value of 5-10 ppb. For the measurements of UFB diameters and number concentrations the NTA is used within all experiments. The observation cell of the Nanoparticle Tracking Analysis (NTA) controls the temperature of each sample to $T = 25$ °C to improve the comparability between the measurements. Prior to every experiment, a sample of the UPW is analyzed by NTA to confirm the purity of the liquid.

3.1.1 Ultrafine bubble generation by pressurized dissolution

For the generation of a large volume of UPW containing UFB, the GaLF FZ1N-10 fine bubble generator by IDEC, Japan, is used. The GaLF is working with the principle of the pressurized dissolution, utilizing the pressure dependency of the gas solubulity in a liquid as explained in section 2.3.2.

As illustrated in figure 3.1a, the fine bubble generator is connected to a basin with a filling volume of $V_{\mathrm{fill}} = 12$ L and creates a closed loop in which the UPW is circulated. A power consumption of 1.5 kW results from the pump during operation. The liquid is pumped with a mean flow rate of $\dot{V}_{\mathrm{L}} = 16.7$ L/min and is continuously fed with air. The ambiant air is self-primed due to the negative pressure caused by the liquid flow (compare figure 2.16). The average air volume flow rate is $\dot{V}_{\mathrm{G}} = 800$ mL/min $\pm 20\%$. A hollow fiber membrane filter at the air inlet retains all particles larger than 10 nm to avoid impurities influencing the UFB measurements.

The GaLF is kept running for six hours to quantify the influence of the operating time on UFB concentration and size. Therefore, samples are taken over time and measured by NTA as described in section 4.1.

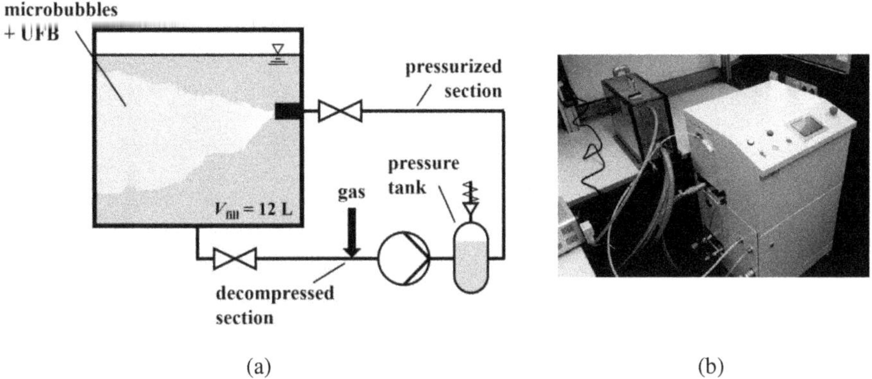

Fig. 3.1 (a) Flow scheme of the setup for ultrafine bubble generation by pressurized dissolution according to [Ter11]. (b) Photo of the setup in the laboratory

3.1.2 Ultrafine bubble generation by ultrasonication

In contrast to most other fine bubble generators, the generation of UFB by ultrasonication comes along without moving parts and reduces a possible contamination of the system to a minimum due to a lack of abrasion. As shown in figure 3.2a, the setup only consists of the liquid sample with the ultrasonic probe dipping inside. The UFB generation with ultrasonication is favorable for small liquid volumes due to the high acoustic pressure required for the cavitation.

For this study, 500 mL of UPW is sonicated using the UP 400s ultrasonication probe by Hielscher Ultrasonics GmbH, Germany (see figure 3.2b). The UP 400s operates with a power of $P = 400$ W and at a frequency of $f = 24$ kHz. Due to the high volumetric power input of $P \cdot V^{-1} = 800$ kW/m^3, the sonication time is limited to 60 minutes to prevent boiling of the UPW. The sonotrode type H14 with a diameter of $d = 14$ mm is used which enables the investigation of different amplitudes A_0 of the ultrasonic waves. To quantify the impact of the amplitude on UFB concentrations and diameters, two different amplitudes $A_0 = 25$ µm and $A_0 = 50$ µm are analyzed. The ultrasonic treatment is pulsed with a pulse length of $t_p = 0.5$ s within a cycle of $t_c = 1$ s.

Using NTA, the change of UFB concentrations and diameters is monitored during the ultrasonication as well as after the generation. The sonicated samples are kept sealed for a week to evaluate the stability of the UFB.

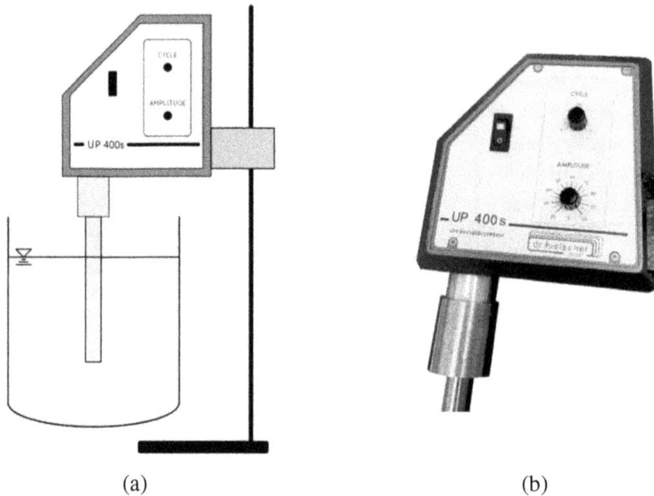

(a) (b)

Fig. 3.2 Sketch of the setup for ultrafine bubble generation by ultrasonic treatment (a) and close-up of the ultrasound generator (b)

3.2 Three reactor setups to evaluate the performance using fine bubble aeration

In biocatalytic and fast chemical reactions, the mass transfer from the gaseous phase into the liquid component is often limited. With regard to process optimizations, the utilization of high-dense finebubble flows promise higher mass transfer rates due to a large volume specific interfacial area of the gaseous phase. To evaluate the impact of fine bubble aeration on the performance of a multiphase contact apparatus, three frequently used reactor concepts are investigated at laboratory scale:

- bubble column reactor
- jet reactor
- stirred tank reactor

Both, hydrodynamics and mass transfer characteristics are analyzed, measuring bubble size distributions (BSD) of the microbubbles as well as determining mass transfer coefficients. The design of the three fine bubble aerated reactor concepts and the experimental procedure for their characterization is described in the following.

3.2.1 Setup of fine bubble aerated bubble column reactor

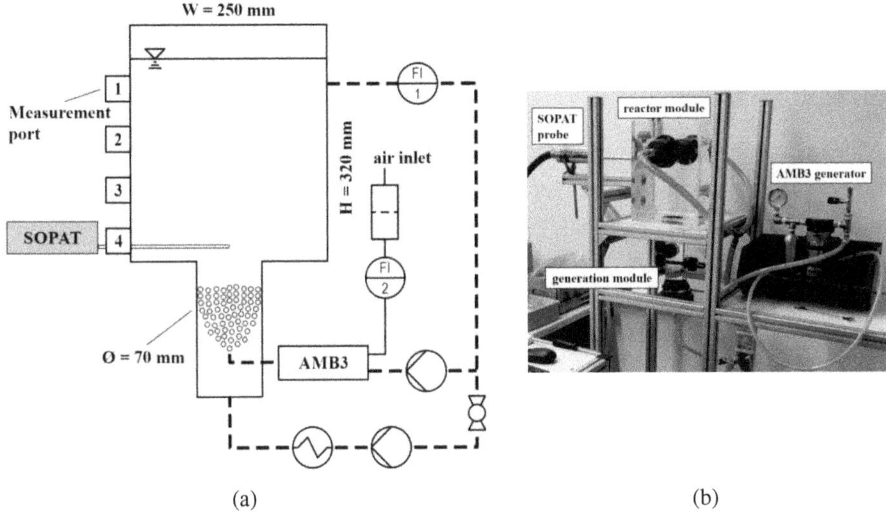

Fig. 3.3 (a) Flow chart of the microbubble aerated bubble column reactor. (b) Photo of the experimental setup

To investigate the hydrodynamics of free rising microbubble flows, a flexible, modular bubble column reactor has been designed. The whole reactor system consists of two parts which are made of acrylic glass to achieve optical accessibility. Figure 3.3 shows schematically the structure of the setup which is devided into the cylindrical microbubble generation module with an inner diameter of $D_R = 70$ mm and an exchangeable reactor module on top of it. For these studies, a cuboid basin (250 x 250 x 320 mm^3) with four measurement ports is used. The measurement ports one to four are equally spaced over the reactor height with a distance of $x = 70$ mm. Through each port an endoscopic probe can be inserted to measure the bubble size distribution of the dense two-phase flow at different positions within the reactor. For the BSD measurements a probe-based microscope (SOPAT Pl) by SOPAT GmbH, Germany, is used. The integrated stroboscope light and the variable reflectors enable to take backlight illuminated pictures within a well defined measurement volume. For the analysis of the data, the SOPAT Sampler and Batcher version 2.1.17.1623 are used. The SOPAT system has an optical resolution down to 3 µm and a maximum frame rate of 15 Hz (all data of the used probes are given in appendix A).

The total filling volume of the reactor is $V_R = 16$ L and for the generation of the microbubbles the AMB3 fine bubble generator by Hack UFB Co., Ltd, Japan, is used, generating microbubbles according to the pressurized dissolution method. The ambient air used for the bubble formation is sucked through a micro pore filter to prevent any contamination influencing the bubble behavior. The overpressure inside the mixing chamber is $p = 0.27$ bar and the volume flow rates are $\dot{V}_L = 1.2$ L/min on the liquid side and $\dot{V}_G = 55$ mL/min for the air flow. For all measurements, deionized water is used and the temperature is kept constant at $T = 25$ °C.

The microbubble generation module provides a homogeneously distributed fine bubble two-phase flow at the inlet of the reactor module. Measurements of the BSD at the reactor inlet over the cross section of the cylindrical generation module confirm their uniformity. In figure 3.4, the cumulative distribution and the characteristic mean diameters d_{50} and d_{32} for the bubbly flow are presented at the center and the edge of the generation module. The cumulative distributions are almost congruent and also the mean diameter shows a good agreement.

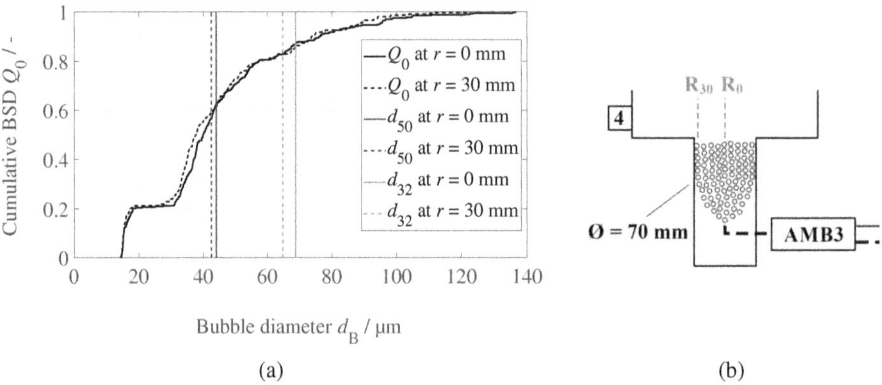

Fig. 3.4 (a) Cumulative BSD measured at the entrance of the bubble column reactor. (b) Illustration of the measurement positions in radial direction

3.2.2 Setup of jet reactor with injector nozzle fine bubble generator

For the characterization of different injector nozzle fine bubble generators a cylindrical glass reactor with a filling volume of $V_{\text{fill}} = 2$ L is used. The filling volume includes all hoses and the pump needed for the injector nozzles. The vessel has an inner diameter of $D_R = 105$ mm, a height of $H_R = 300$ mm and a double jacket to keep the liquid temperature costant at $T = 25$ °C during all experiments. The fine bubble generator is installed in the center of the reactor lid. The nozzle is vertically oriented with its outlet at a height of $h = 200$ mm over the reactor bottom. For the operation of the fine bubble generator a liquid flow is provided by a gear pump which is connected to an outlet at the reactor bottom. Pressurized air is used for the aeration and the gassing rate is held constant by a flow controller (Bronkhorst L-Flow).

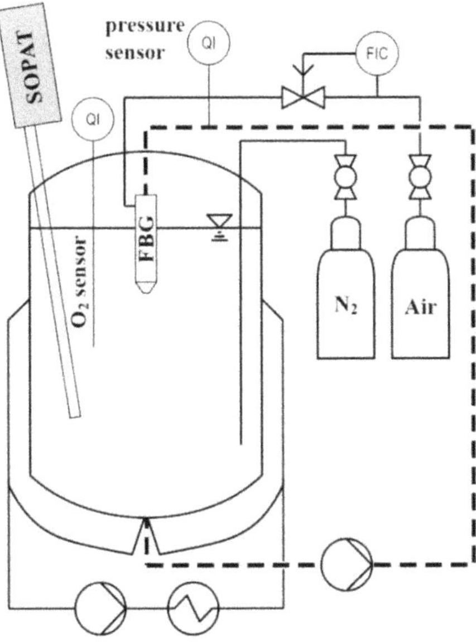

Fig. 3.5 Flow chart of the microbubble aerated jet reactor

The pressure drop Δp caused by the fine bubble generator is measured to determine the power which is supplied to the system by the gear pump. The power input of a two-phase nozzle

$$P = \dot{V}_L \cdot \Delta p \tag{3.1}$$

is calculated by the liquid volume flow rate $\dot{V}_L$ and the measured pressure drop Δp. Due to the liquid flow structure within the nozzles, a negative pressure is created sucking in the gaseous phase. Therefore, the influence of the gaseous phase on the power input is neglectable in that case. Two different fine bubble generators are investigated and compared to each other with regard to the produced bubble sizes and their mass transfer performance. Figure 3.6 shows both fine bubble generators. The Awataro A-PW-03 by Nitta, Japan (figure 3.6a), works with the swirl principle whereas the Schlick series 803 by Düsen Schlick GmbH, Germany (figure 3.6b) belongs to the group of ejector type fine bubble generators.

The influence of the gas flow rate as well as the influence of the power input on the performance of the generators is analyzed. The gas flow rate is varied from $\dot{V}_G = 20$ -

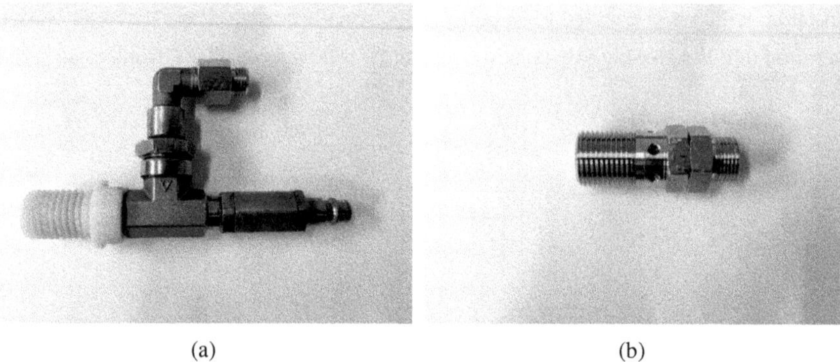

(a) (b)

Fig. 3.6 Photo of the experimental analyzed injector nozzle fine bubble generators. (a) Swirl type generator (b) Ejector type generator

120 mL/min and three different volumetric power inputs $P \cdot V^{-1} \in \{6, 8, 10\}$ kW/m^3 are investigated. The given power inputs are set by the minimum required liquid flow rates for an operation of the two fine bubble generators. For all experiments, the BSD are measured using the endoscopic probe SOPAT Sc (see appendix A). For an evaluation of the mass transfer performance of both fine bubble generators, the volumetric mass transfer coefficient $k_{\mathrm{L}}a$ is determined by the dynamic method. Therefore, the liquid phase is aerated with nitrogen to strip out the dissolved oxygen until a concentration lower than 10% of the atmospheric oxygen saturation level is reached. Subsequently, the fine bubble aeration with air is started and the increase of the dissolved oxygen concentration c_{O_2} is recorded [Ros19]. The $k_{\mathrm{L}}a$ value is calculated according to

$$k_{\mathrm{L}}a = \frac{dc_{\mathrm{O}_2}}{dt} \cdot \frac{1}{(c^* - c_{\mathrm{O}_2})} \tag{3.2}$$

with the saturation concentration c^* and the measured oxygen concentration c_{O_2} at its corresponding time t. For the measurement of the dissolved oxygen concentration, the optical oxygen probe FDO 925 by WTW, Germany, is used. The stripping with nitrogen to remove the dissolved oxygen from the liquid is a well established procedure in process engineering. Due to its easy handling and low costs, nitrogen is mostly used as stripping gas. The influence of the stripping gas on the mass transfer will be discussed later on (see section 5.4.3). For the comparison of the different reactor systems with regard to measured mass transfer coefficients, nitrogen is used following the standard procedure.

3.2.3 Setup of fine bubble aerated stirred tank reactor

The application of fine bubble aeration to stirred tank reactors is evaluated using two laboratory scale glass reactor setups. The 3 L STR with a diameter of $D_R = 130$ mm has a rounded bottom whereas the 2 L STR with a diameter of $D_R = 105$ mm comes along with a flat bottom. Both reactors have a height of $H_R = 300$ mm and are equipped with three baffles. Equally to the jet reactor setup, the temperature is controlled to $T = 25$ °C using a double-jacket. To reduce the friction and to ensure optimal stability, the stirrer shaft is fixed in the center of the reactor lid using ceramic ball-bearings. The stirrer is driven by a ViscoPakt-rheo - X7 by HiTec Zang GmbH, Germany, which additionally measures the afforded torque with a resolution of $M_{res} = 0.003$ Ncm and a reproducibility of $M_{rep} = 0.05$ Ncm. For the agitation different stirrer configurations with different impeller types are used. Single-stage stirrers are investigated, comparing the Rushton turbine ($d_S = 54$ mm), a pitched blade impeller ($d_S = 54$ mm) and a segment impeller ($d_S = 54$ mm) with regard to their influence on the fine bubble aerated system. The impellers are mounted at a height of $h/d_S = 1$.

A three-stage configuration consisting of two Rushton turbines mounted at the bottom and a pitched blade impeller at the top is analyzed as well to enable a comparison to the jet reactor system with regard to the induced power. For the multi-stage stirrer, all impellers are mounted equally distributed with a spacing of $s/d_S = 1$.

The fine bubble aeration is realized following two different approaches. On the one hand, the AMB3 fine bubble generator with an additional flow circuit is used to analyze the hydrodynamics of the system at the highest fine bubble densities. On the other hand, static fine bubble aeration is applied by using porous spargers installed bellow the stirrer. Therefore sintered frits and SPG membranes with different porosities from 0.5 µm to 2.0 µm are utilized. Additionally, the gassing by an open tube with an orifice of $d_O = 4$ mm is investigated to ensure comparability to conventionally aerated STR systems.

As visualized in figure 3.7, BSD are measured optically inside the bubbly flow using the SOPAT Pl and Sc probe (see appendix A) and also from outside of the reactor. In this case, pictures are taken with an Optronis high-speed camera with a resolution of 450 x 450 pixels. For the illumination a LED panel is installed behind the reactor, at the opposite of the camera. To compensate the optical distortion due to the curvature of the reactor, it is placed inside a cuboid basin.

The determination of volumetric mass transfer coefficients by the dynamic method is done similar to the jet reactor setup (compare chapter 3.2.2) and according to equation 3.2. Besides nitrogen, also argon and helium are used for the stripping to analyze the influence of the stripping gas on the measured mass transfer coefficients. Furthermore, the liquid phase is degased physically by boiling it for several minutes to drive out all dissolved gases. In an airtight container the medium is cooled down to the operating temperature of $T = 25$ °C before it is decanted back to the reactor.

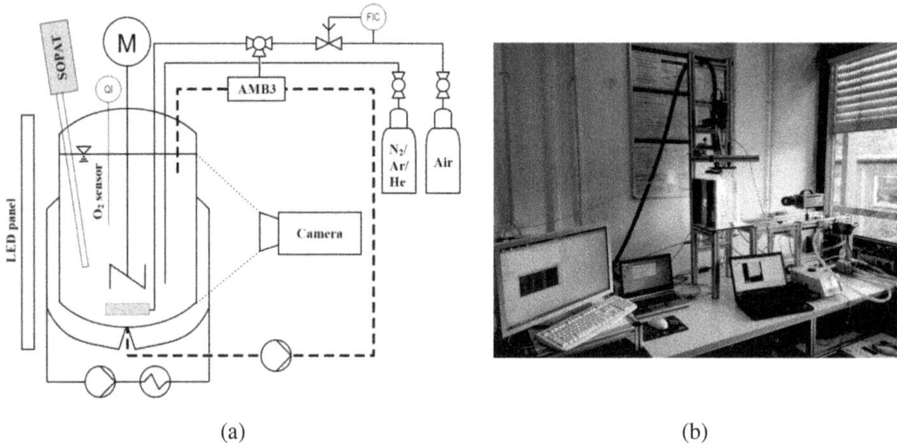

Fig. 3.7 (a) Flow chart of the microbubble aerated stirred tank reactor. (b) Photo of the experimental setup

3.2.4 Liquid systems for fine bubble aerated reactor setups

The limitation of the mass transfer in two-phase systems is a crucial aspect for process optimization in various fields of process engineering. For example enzymatic oxidations in aqueous solutions require a sufficient oxygen supply and are often limited due to the low solubility of oxygen within the liquid phase. In the context of this work a joint project with the Institute of Technical Biocatalysis (ITB) at the TU Hamburg has been carried out, analyzing the impact of fine bubble aeration on enzymatic reactions. The ITB has investigated the benefit of fine bubble aeration for biocatalysis using the enzymatic sodium gluconate synthesis as model reaction [Tho20].

The hydrodynamics and mass transfer characteristics presented within this work are determined for three different liquid systems. The media and their properties are given in table 3.1. For a better transferability of results within the joint project with the ITB, a glucose/BSA system has been used in the experiments. The mixture of deionized water with glucose and the standard protein bovine serum albumin (BSA) is a lower-cost and longterm stable model system for the enzymatic sodium gluconate synthesis. Deionized water is used due to its easy handling and the large amount of literature data available for comparison of results. The frequently used standard surfactant Triton X-100 is added to the

DI water achieving reduced surface tensions as present in real reaction media and to analyze the influence of surface active substances on fine bubble aeration.

Table 3.1 Analyzed liquid systems and their surface tension measured at $T = 25$ °C

Liquid system	Surface tension / $mN \cdot m^{-1}$
Deionized water	71.0 ± 0.1
DI water + glucose (133 mmol/L) + BSA (67 mg/L)	63.2 ± 0.1
DI water + Triton X-100 (0.24 mmol/L)	39.2 ± 0.1

3.3 Setups for investigating local mass transfer effects at microscale bubbles

In mass transfer limited processes, the optimization of transferring a certain species from the gaseous to the liquid phase is still challenging. The aeration with microscopic bubbles promises high mass transfer rates and represents a possibility to overcome the mass transfer limitation. A decisive role is played by the large volume specific surface area of the bubble and the no longer negligible influence of the Laplace pressure.

For a better understanding of the acting mechanisms during the shrinking of a microbubble, a setup for the generation of size adjustable single microscale bubbles is developed. Using two different experimental approaches, local concentration fields are measured utilizing the LIF method as described in section 4.2.

3.3.1 Setup for generating single microscale bubbles

For the generation of single bubbles with diameters of several micrometers, a fused silica capillary tubing by Molex, LLC, USA, is used. The capillary has an inner diameter of $d_{in} = 2$ µm and an outer diameter of $d_{out} = 150$ µm and is located in the center of the measurement cell. Due to its coating, the capillary is characterized by a flexibility which allows non-destructive bending. This flexibility is utilized to remove the bubble at its desired size from the capillary. Due to the acting surface tension and the low buoyancy force, an external trigger is needed to detach the bubble.

As illustrated in figure 3.8, the external force is applied by a plunger which is connected to a converted solenoid valve. The magnetic field induced by the electric coil pulls back the plunger so that the spring is tensioned. When the power is turned off, the plunger is pulled forward by the spring.

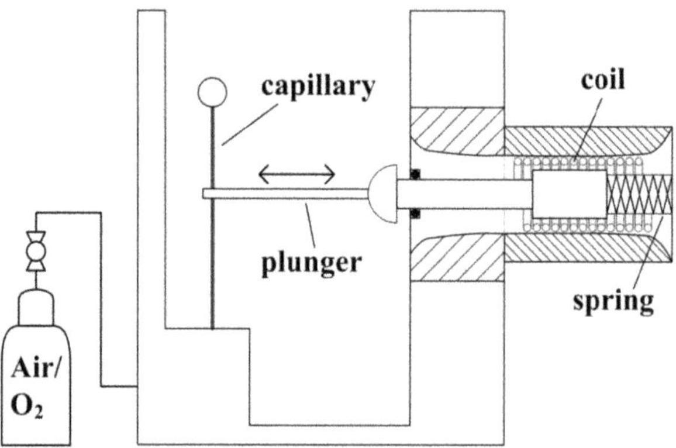

Fig. 3.8 Illustration of the working principle for the generation of a single microscale bubble

The magnetic plunger is controlled by an in-house designed Arduino based syncronizer. Figure 3.9 shows the detachment process recorded by a pco.dimax HS2 highspeed camera with a resolution of 1400 x 1050 pixels and a frame rate of 1000 fps. As lens the CentriTel Focuser Model K2 with an optical magnification of 3.56 times is used. The bubble detaches from the capillary after less than 0.02 s.

Due to the small cross-section area of the capillary, only a small liquid volume gets displaced during the bubble detachment procedure. The small induced liquid flow has no influnce on the rising behavior of the microscopic bubble which is rising straight to the top of the measurement cell as visible in figure 3.9. The measurement cell has a depth of 4 mm ensuring a large distance between the rising bubble and the walls of the cell. Therefore, wall effects influencing the rising of the bubbles, as described by [Cli78] are neglected for the analyzed microscopic bubbles.

3.3.2 Setup for measuring local concentration fields at a free rising microscale bubble

For the measurement of concentration fields in the vicinity of a rising microscale bubble by planar LIF, the above described setup for the generation of single microscale bubbles is expanded by a high power LED (OSRAM OSTAR Projection Power) with a wavelength of $\lambda = 459$ nm and a self-developed light sheet optics.

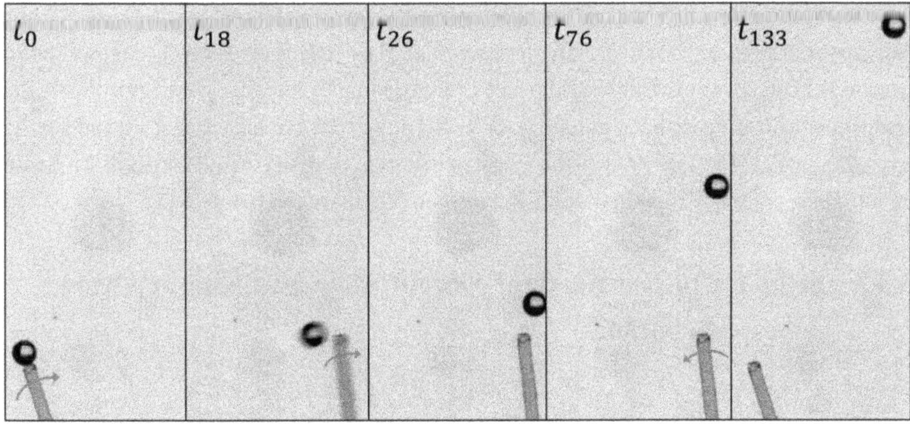

Fig. 3.9 Visualization of the detachment process of a microscopic bubble ($d_B \approx 280$ µm) from the capillary

The light sheet optics consists of a focus lens and two shutters positioned in a row. A light sheet with a constant thickness of 1 mm is produced over the entire width of the measurement cell. As illustrated in figure 3.10, the light sheet and the camera are arranged perpendicular to each other to avoid any reflections.

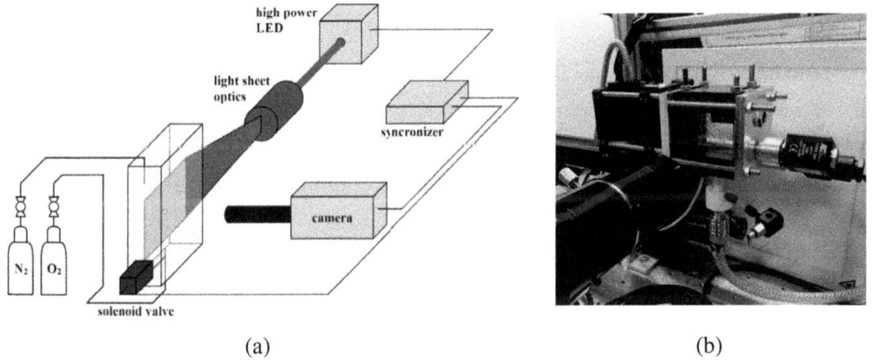

Fig. 3.10 (a) Scheme of the experimental setup for p-LIF measurements at microscale bubbles. (b) Photo of the experimental setup for p-LIF measurements at microscale bubbles

Within these investigations, pure oxygen is used for the bubble formation in deionized water. The fluorophore dichlorotris(1, 10-phenanthroline)ruthenium(II)hydrate is added to the deionized water with a concentration of 30 mg/L. The fluorescence dye has its maximum absorption rate at the wavelength of the used LED and primarily emits light with a wavelength of $\lambda = 590$ nm. A bandpass filter (center wavelength $\lambda = 590$ nm ± 2 nm, half-power bandwidth 20 nm ± 2 nm, transmission $> 84\%$ by ILA_5150 GmbH) is used in front of the camera to exclude the influence of diffused light. Pictures are recorded with 500 fps and the LED is operated in pulsed mode with a pulse width smaller than 2 ms and a pulse repetition rate of 0.5 kHz. Prior to the measurements the oxygen is stripped out of the liquid by nitrogen aeration to an initial dissolved oxygen concentration below 10%.

3.3.3 Setup for measuring local concentration gradients at a fixed microscale bubble

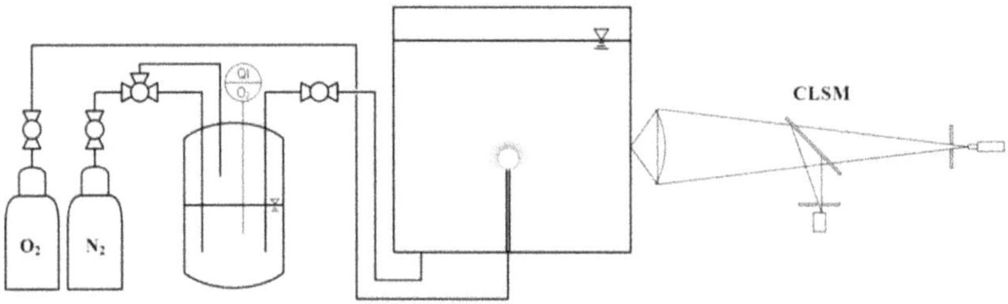

Fig. 3.11 Scheme of the setup for investigations of local concentration gradients at a fixed microscale bubble

For the measurement of concentration fields with a high optical resolution on the microscale, the confocal laser scanning microscopy is used. The pictures are taken using an Olympus Fluoview 1000 CLSM equipped with the UPLFLN10X2 lens and a wavelength of $\lambda = 458$ nm. The working distance of 10 mm in combination with a 10x magnification allow a spatial resolution of around 1 µm per pixel. Due to a specially designed wire rope suspension, the used CLSM system is tilted by 90 degrees which allows observations from a lateral viewing direction. For the investigation of local dissolved oxygen gradients, a fixed microscopic oxygen bubble in nitrogen stripped deionized water is analyzed. As for the planar LIF measurements, dichlorotris(1, 10-phenanthroline)ruthenium(II)hydrate is added to the deionized water with a concentration of 30 mg/L to visualize the dissolved oxygen. An observation area of 64 x 640 pixels is scanned to capture the middle section of the bubble and the surrounding liquid. For a sufficient image quality, an exposure time of 25 µs per pixel is required resulting in a maximum recording frequency of 1 Hz for the investigated region of interest. The limitation in the scanning rate does not allow the investigation of a moving, free rising microbubble. As shown in figure 3.11, the fluorophore solution is aerated with nitrogen to an initial dissolved oxygen concentration close to zero before it is fed into the measurement cell. All the tubing is realized gas tight to prevent oxygen from the air diffusing into the system. The measurement cell is made out of a 2 mm flat and 50 mm wide glass capillary with the fused silica capillary integrated in its center. To ensure that only pure oxygen bubbles are generated, the fused silica capillary and all connected hoses are flushed with oxygen for several minutes prior to the measurement.

A PreSens Needle-Type Oxygen Microsensor (response time < 3 s) is installed to the reservoir of the fluorophore solution to monitor the dissolved oxygen concentration of the feed. To obtain quantitative information on the oxygen concentration fields, a calibration is done for every measurement. Therefore, four different dissolved oxygen concentrations between zero and maximum saturation are set inside the reservoir and then fed into the measurement cell. As exemplarily shown in figure 3.12, for each of the 64 x 640 pixels a calibration curve is defined assigning an oxygen concentration to a grey value according to the approach of Stern-Volmer [Tim18].

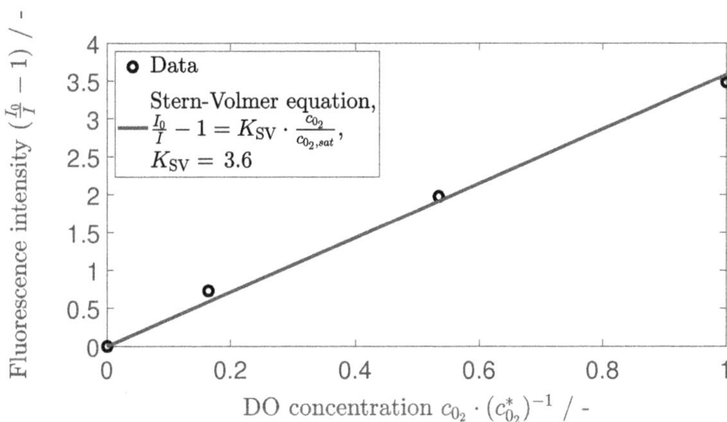

Fig. 3.12 Example for the calibration during the CLSM measurements according to Stern-Volmer

Chapter 4

Measurement techniques

The investigation of ultrafine bubbles at nanoscopic scale as well as the characterization of fine bubble aerated two-phase flows pose a particular challenge to measurement technology due to the small dimensions. With regard to hydrodynamics and mass transfer performances of fine bubble aerated systems, the determination of the gas hold-up as a crucial process parameter and the visualization of concentration fields in the surrounding of microscopic bubbles are challenging. This chapter provides a description of special measurement techniques that are used within the scope of this work:

- Nanoparticle Tracking Analysis
- laser induced fluorescence
- confocal laser scanning microscopy
- gas hold-up determination from endoscopic measurements

First, the Nanoparticle Tracking Analysis (NTA) is explained which enables an insight into nanoscale systems measuring particle sizes and concentrations down to 12 nm. Then, the techniques of laser induced fluorescence (LIF) and confocal laser scanning microscopy (CLSM) for the visualization of concentration fields in multiphase flows with a high spatial resolution are described. Finally, an approach for calculating gas hold-ups in homogeneous fine bubble flows is presented.

4.1 Nanoparticle Tracking Analysis

Nowadays the number of nanomaterial based consumer products is strongly increasing. With regard to their impact on the environment suitable methods for characterizing those systems are needed. In multiphase systems the measurement of particle sizes and concentrations of a dispersed phase at nanometer scale is still challenging. Conventional optical measurement techniques are limited by the wavelength range visible to the human eye from $\lambda = 400 - 700$ nm. Utilizing the mechanism of light being scattered at the surface of small particles, the Nanoparticle Tracking Analysis enables the measurement of the size distribution and concentration in dispersed systems with a spatial resolution down to 12 nm. Monodispersed as well as polydispersed nanoparticles are accurately characterized by NTA making it applicable for all types of nanoparticles [Fil10].

4.1.1 Principle and equipment

The NTA combines the properties of both light scattering as well as Brownian motion to determine size distribution and concentration of nanoparticles in a dispersed liquid system. The main parts of a NTA setup as displayed in figure 4.1 are the laser beam (wavelengths from $\lambda = 405 - 635$ nm are mostly used), the sample chamber and a conventional optical microscope equipped with a complementary metal–oxide–semiconductor (CMOS) camera.

The laser beam enters the measurement chamber through a prism-edged glass flat. The laser beam gets refracted at the interface of the glass and the liquid sample layer leading to a reduced profile with simultaneous high power density [Hol13]. Particles which are passing through this compressed laser beam scatter the light so that it is visualized by a microscope objective with a long working distance and 20x magnification. The CMOS camera is recording a video for several seconds where the movement of the scattered light is captured. The calculation of the particle diameters is done by the NTA software. This can lead to small deviations between different commercial systems or software versions and should therefore always be specified. All NTA results presented in this work are analyzed using NTA Version: NTA 3.2 Dev Build 3.2.16. The individual steps for the calculation are shown in figure 4.1.

The software marks the center of each captured particle and tracks its movement under Brownian motion for each frame of the recorded video. From this information the averaged distanced traveled in x- and y-direction is calculated for each particle. Having these information, the hydrodynamic diameter χ of the nanoparticles is calculated using the Stokes-Einstein equation [Ein56]

$$\frac{\overline{(x,y)^2}}{4t} = D = \frac{kT}{3\pi\eta\chi} \tag{4.1}$$

where D is the diffusion coefficient, and by knowing the time t between the traveled distance, the Boltzmann constant k, the Temperature T and the dynamic viscosity of the liquid η.

The accuracy of the measuring system is increased by controlling the temperature of the measurement chamber to a constant value.

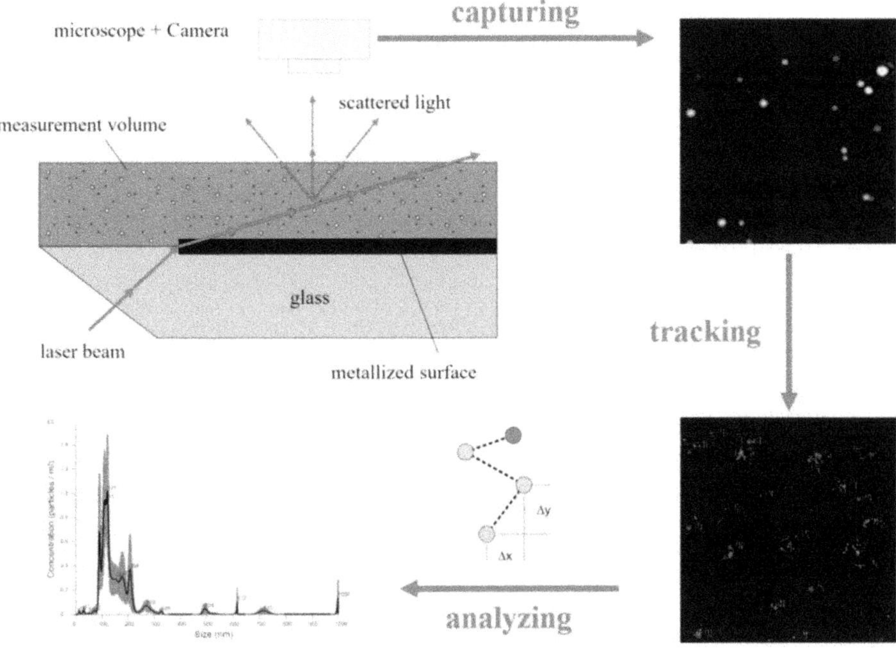

Fig. 4.1 Scheme of the working principle of NTA and illustration of the single steps in the post processing

4.1.2 Application to ultrafine bubble liquids

Most available finebubble generators produce UFB within a wide range of a several hundred nanometers. Due to the gaseous state of UFB their size is not only affected by the generation principle but also by the required local thermodynamic equilibrium at the gas-liquid interphase. Furthermore, coalescence effects can occur which lead to a broad distribution of the UFB size. Therefore, UFB systems are mainly polydispersed what makes their characterization a challenging task. The fact that NTA is, compared to other nanoparticle measurement techniques such as Dynamic Light Scattering, not biased towards larger particles or agglomerations makes it a suitable method for the characterization of UFB solutions [Hol13]. A prerequisite for the application of NTA for UFB is a neglection of buoyancy forces so that the movements of the captured particles are dominated by omnidirectional Brownian motion [Bod17].

A limitation of the NTA is that it is not possible to distinguish between nanoparticles by their aggregate state. A gaseous UFB scatters the light in the same manner as a solid nanoparticle. Therefore, applying NTA in the field of UFB requires a very clean working environment and an experimental procedure which helps to exclude a contamination by solid nanoparticles. In the last ten years, several research groups have successfully used NTA for the characterization of UFB solutions by following different experimental approaches applying supplement measurement techniques such as freeze-fracture electron microscopy [Ohg10, Uch11, Wan19].

4.2 Laser induced fluorescence

For a better understanding of the mass transport phenomenology in mutliphase flows, the laser induced fluorescence imaging technique is used for the visualization and quantitave characterization of mass transfer effects. With a high spatial and temporal resolution, instantaneous concentration fields of time dependent processes are recorded [Rüt18]. Especially at the interface of two phases the LIF offers local insights into the concentration boundary layer. Within this work, LIF is used for the characterization of a gas-liquid system with oxygen bubbles as dispersed phase. A more detailed description on LIF with further information regarding different fields of applications as well as opportunities and draw-backs of this technique is given in the review of Rüttinger et al. [Rüt18].

The principle of LIF is based on the sensitivity of a fluorescent dye to the transferred species of the considered process. The fluorescent dye is added to the liquid bulk phase as tracer and is excited by laser light. Its fluorescence intensity is captured by high-speed imaging. Depending on the dissolved oxygen concentration in the liquid, the fluorescence intensity is changing. To get quantitative information from the measurement a calibration function has to be found which gives a correlation between the dissolved oxygen concentration and the grey levels in the recorded image. Therefore, fluorescent dye solutions with different known oxygen concentrations have to be measured prior to the experiment.

For a high sensitivity of the measurement, the excitation wavelength of the laser has to match the absoprtion maximum of the fluorescent dye. To achieve an optimal measurement result, the dye should be soluble in the bulk phase and should not change the fluid properties. Furthermore, the dye has to be stable under the exposed laser intensity and its absorption and emission spectra have to be separate from each other [Kar97]. Figure 4.2 displays the absorption and emission spectrum of dichlorotris(1, 10-phenanthroline)ruthenium(II)hydrate which is used for the visualization of dissolved oxygen concentration fields. The small overlapping and the clearly separated intensity maxima make it ideal for LIF measurements.

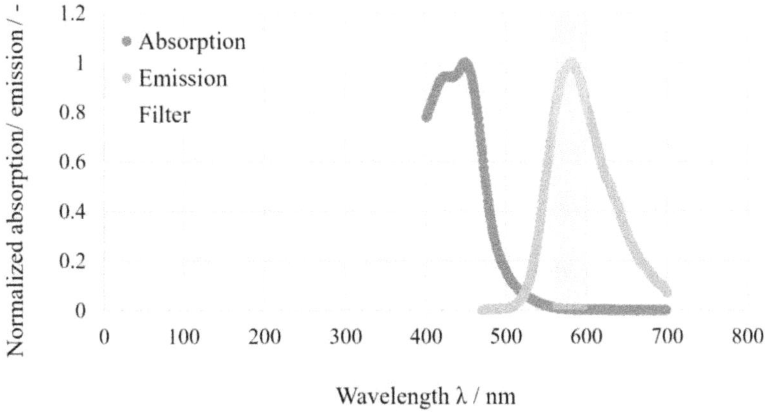

Fig. 4.2 Absorption and emission spectrum for the used ruthenium complex dichlorotris(1, 10-phenanthroline)ruthenium(II)hydrate

By using a bandpass filter (compare figure 4.2) for the recording of the images, the camera is secured for direct illumination by the laser and only the emitted fluorescence light is recorded, enhancing the contrast of the images. The spatial and temporal resolution of a LIF imaging system is defined by the camera and the used lens as well as the illumination setup. In contrast to volume illumination, more detailed results are optained by illuminating just a thin slice using light sheet optics. The highest spatial resolutions are achieved by a combination of LIF and confocal laser scanning microscopy which is described in the following.

4.3 Confocal laser scanning microscopy

The analysis of microbubbles using conventional widefield laser-optical measurement techniques can state problems due to the spatial expansion of the used light source. Having a laser beam being thicker than the investigated region of interest results in low-contrast images caused by bright signals outside the focus leading to an overshining of the focus plane. To overcome this limitation, confocal laser scanning microscopy is used to obtain highly resoluted fluorescent images at micro scale.

The CLSM is similar to conventional fluorescent microscopy in addition with a confocal scan head. The scan head itself has inputs from one or more external laser light sources and is equipped with fluorescence filter sets. In addition to that, its raster scanning mechanism is based on a galvanometer and enables a 2D-scanning within the focus plane. By using one or more variable pinhole apertures, the confocal image is generated. In order to analyse different fluorescent wavelengths, the scan head has built-in photomultiplier tube (PMT) detectors (compare figure 4.3).

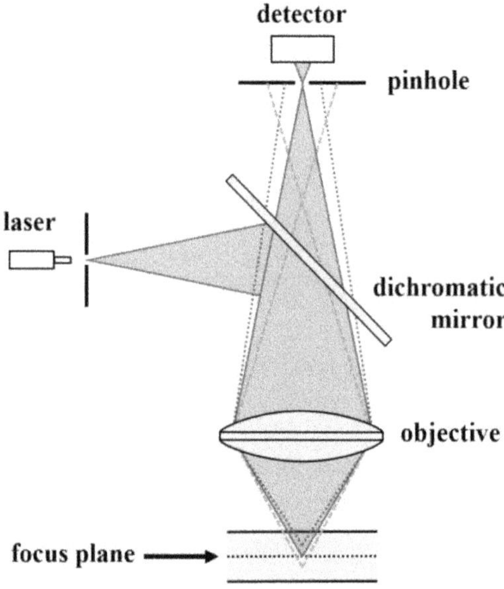

Fig. 4.3 Illustration of the working principle of a CLSM scan head according to [Mur13]

The CLSM makes use of the basic principle of epi-illumination, in which the light source and the detector are located on the same side of the sample and separated from each other by the optical arrangement. In order to form an intense diffraction-limited spot inside a fixed focus plane, the laser beam is expanded to fill the rear aperture of the objective [Mur13]. The scanning proceeds from side to side and from top to bottom over the sample in a raster-like pattern by deflecting the laser beam. Each scanned point corresponds to one pixel in the final image. Changing the focus of the lens using a stepper motor, 3D information are obtained as well.

The key component of the CLSM is the pinhole aperture which largely excludes fluorescence signals from objects above and below the focus plane (illustrated as dashed lines in figure 4.3). Fluorescence signals from out-of-focus objects are described as diffuse extended disks so that only a negligible small fraction will pass the pinhole. Furthermore, the pinhole reduces the stray light inside the optical system. This ensures that only fluorescence information within the focus point of the laser beam passes the pinhole aperture to get detected. The PMT detector produces a voltage signal that corresponds to the intensity of incident fluorescent photons. By digitizing those signals an image is generated. The restriction of the CLSM is given by its low temporal resolution which makes it mainly used for stationary processes.

4.4 Determination of the gas hold-up from endoscopic measurements

For an exact and complete description of a dispersed system the hold-up of the gaseous phase is required. With regard to mass transfer characteristics the total amount of gas within the liquid phase that is participating in the reaction is of importance. Knowledge of the gas hold-up in combination with the bubble size distribution enables the calculation of the volume specific interfacial surface area a (according to equation 2.26) and therefore the determination of the mass transfer coefficient k_L.

The measurement of the gas hold-up is a challenging task in process industry. An often applied method is the measurement of the change in the liquid height due to the aeration. The uncertainty of this method is given by the fluctuations of the liquid surface what makes it more likely applicable for high gas hold-ups.

In fine bubble aerated systems gas hold-ups smaller than 1% are typical what makes an observation of changes in the liquid height impossible. For a fine bubble aerated stirred reactor system the gaseous phase is homogeneously distributed and the bubble size distribution is nearly constant over the reactor height [Dru15]. Therefore, there are only slight differences between the overall gas hold-up and localy measured values [Mat20]. By using this circumstance, the gas hold-up for the fine bubble aerated STR setup is calculated with the data obtained by the SOPAT probe. As shown in the figures 4.4 and 4.5, the SOPAT probe has a well defined measurement volume V_{SOPAT}. From all taken pictures ($N_{pictures}$) the total volume of the detected bubbles V_B inside the focus level of the camera is calcualted. Using these values the gas hold-up is defined as

$$\varepsilon_G = \frac{V_B}{V_{SOPAT} \cdot N_{pictures}} \cdot \chi_{ratio} \tag{4.2}$$

with χ_{ratio} being the ratio of all bubbles (number of undetected bubbles outside the focus plus number of detected bubbles inside the focus) to the number of detected bubbles. The value of χ_{ratio} has to be determined manually for each measurement. The basis of equation 4.2 is that all undetected bubbles are statistically equally distributed as all detected bubbles.

Fig. 4.4 Photo of the SOPAT probe within the STR setup and analyzed picture of bubbles using the SOPAT Sampler and Batcher version 2.1.17.1623. For the given picture the correction factor is determined as $\chi_{\text{ratio}} = 27/8$

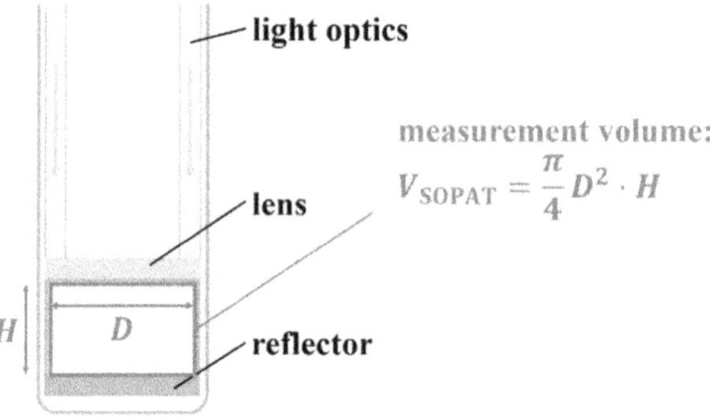

Fig. 4.5 Illustration of the cylindrical measurement volume of the SOPAT probe with the reflector gap H and the diameter of the field of view D.

Chapter 5

Experimental results and discussion

To clarify the potential of using fine bubbles in mass transfer limited processes, the fundamentals on their hydrodyanmics and mass transfer behavior are investigated using the setups explained in chapter 3. The results are presented from smallest to largest bubble sizes and with rising level of complexity. They are divided into four main parts:

- proof of stability for ultrafine bubbles in aqueous liquids
- evaluation and comparison of fine bubble aerated reactor concepts
- characterization of the membrane aerated STR concept
- fundamental mass transfer characteristics of microscale bubbles

In the first part, a proof of stability for UFB is given by analyzing two different generation techniques and a conclusion is drawn on their application in process industry.

The second part focuses on the impact of fine bubble aeration on bubble column, jet and stirred tank reactor systems. BSD and mass transfer coefficients are measured and the fine bubble aerated jet and stirred tank reactors are compared with regard to their efficiency.

In the third part, the membrane aerated STR is further characterized investigating the influence of different process parameters, such as power input or gassing rate, on the hydrodynamics and the mass transfer performance of the system.

At the end, fundamental mass transfer characteristics of microscale bubbles are analyzed for both, single bubbles as well as dense fine bubble flows. The focus here is on the influence of counterdiffusion effects on the mass transfer.

5.1 Proof of stability for ultrafine bubbles in aqueous liquids

For the analysis of ultrafine bubble characteristics, the generation principle and its influence on the material system have to be taken into account. Therefore, the evaluation of particle size distributions measured by NTA is monitored during the generation process. A further discussed topic is the thermodynamic equilibrium between ultrafine bubbles and the bulk phase resulting in a long-term stability. This effect has been studied observing UFB concentrations and diameters of generated samples over one week. Summarizing the results, a conclusion on the role of ultrafine bubbles for process engineering is drawn.

5.1.1 Evaluation of particle size distribution

The generation of ultrafine bubbles is a continuous process in which the running time of the generator influences the characteristics of the generated bubbles. Figure 5.1 shows the evolution of the UFB number concentration and the mean UFB diameter for the pressurized dissolution generator setup. These fine bubble generators are working at their optimum operating point with fixed gas and liquid volume flow rates as well as a constant overpressure inside the mixing tank. With increasing running time of the generator, the UFB concentration inside the system is also increasing. After a strong increase of the concentration at the beginning, a saturation level at around $2 \cdot 10^8$ particles/mL is reached after several hours (compare figure 5.1a).

The UFB concentration is determined by the amount of gas dissolved in the mixing tank and therefore directly correlated to the overpressure. Due to the closed loop system an equilibrium between the number of newly generated bubbles and of those which are removed by the pump is given. The time until the equlibrium state is reached depends on the total liquid volume within the system. Bubble coalescence or break-up do not have an influence on the UFB concentration as indicated by a constant mean UFB diameter ($d_{\mathrm{B}} \approx 150$ nm) during the generation (see figure 5.1b). For a further increase of the maximum UFB concentration the dissolution process of gas under pressure has to be optimized.

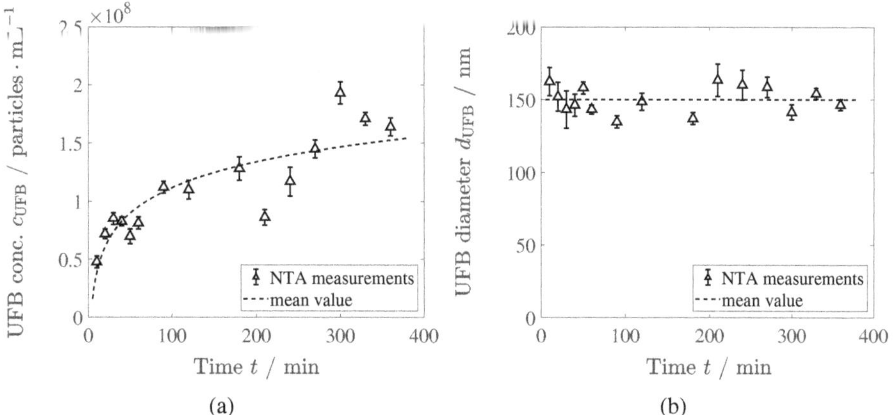

Fig. 5.1 (a) Development of the UFB concentration during running time of the pressurized dissolution fine bubble generator. (b) Deviation of the corresponding mean UFB diameter over time

In contrast to the pressurized dissolution method, the UFB generation by ultrasonication takes place as a batch process. Due to the smaller liquid volume of 500 mL, the running time is reduced to 60 minutes. For an amplitude of $A_0 = 50$ µm, the UFB concentration profile is similar to the one obtained by the pressurized dissolution setup. After 30 minutes, the concentration values start fluctuating around a constant level in the same magnitude as the saturation concentration of the pressurized dissolution generator. For an amplitude of $A_0 = 25$ µm, the measurement shows in the beginning a rapid increase of the UFB concentration with a maximum of $2.5 \cdot 10^8$ particles/mL after 20 minutes of sonication. After this time a further sonication with ultrasound leads to a slow decrease of the concentration down to values below $1 \cdot 10^8$ particles/mL after 60 minutes. As for the pressurized dissolution method, the UFB diameters are less affected during the generation process with a mean UFB diameter of $d_B = 130.97 \pm 13.14$ nm for the 25 µm amplitude and $d_B = 148.32 \pm 24.41$ nm for the 50 µm amplitude.

The UFB formation using ultrasound is based on cavitation effects due to local pressure fluctuations induced by the sonic waves. Reflections at the walls of the vessel are leading to a complex pressure distribution inside the liquid with local deviations. Therefore, large measuring errors within the results are traced back to the manual sample extraction for the NTA resulting in slight variatons of the sampling location. The pressure fluctuations can also cause bubble bursting explaining the decrease in the UFB concentration.

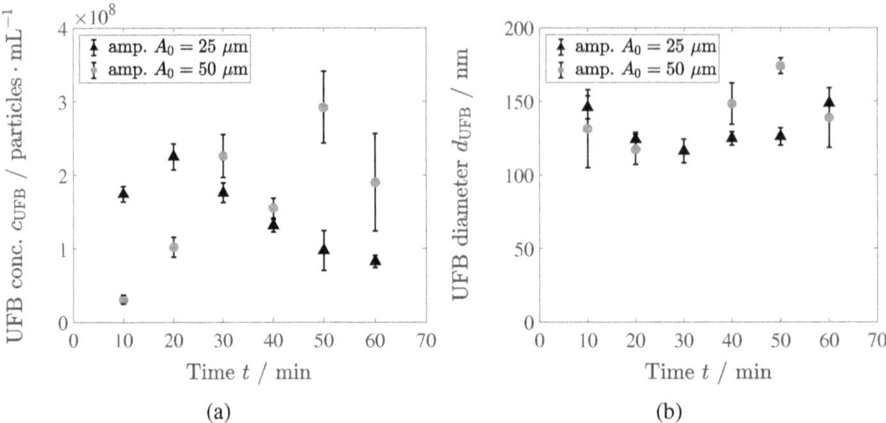

Fig. 5.2 (a) Development of the UFB concentration during sonication for two different ultrasound amplitudes. (b) Deviation of the corresponding mean UFB diameter over time

For the evaluation of the stability of UFB, sonicated samples have been kept sealed for one week and changes in the UFB concentrations and diameters are observed. For both amplitudes a similar behavior is recognizable. After seven days there are still strong signals detectable by the NTA inside the samples. Compared to the initial measurements the concentrations have been decreased strongly with a simultaneous increase of the mean diameter. Wang et al. [Wan19] observed a similar behavior for UFB generated by periodic pressure change 48 hours after their generation.

Assuming a gaseous state of the particles, the decreasing concentration is caused by mass transfer from the UFB into the liquid phase. Local supersaturation gets reduced over time by diffusion so that complete dissolution of bubbles can occur. The increasing UFB diameter can be explained either by coalescence of UFB what would also reduce the UFB concentration or by bubble growth due to the transfer of dissolved gas into the bubble. In what percentage which of these mechanisms take place requires deeper investigations being beyond the scope of this first experimental approach on proving UFB existence.

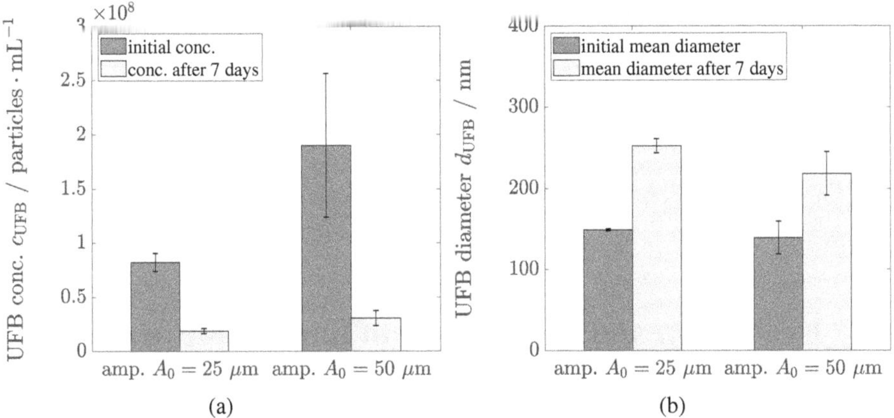

Fig. 5.3 Change in the UFB concentration (a) and mean diameter (b) during storage for one week

5.1.2 Conclusion on the existence of ultrafine bubbles and their application to process industry

The previous results on the generation of air UFB in ultrapure water shows a maximum achievable UFB concentration in the magnitude of 10^8 particles/mL independent on the applied generation principle. Also the size distribution of the UFB is similar for both setups showing only slight differences in their mean diameter. For a clean system, bulk UFB can be stabilized by a diffusive shielding effect inside a bubble cluster as described by [Wei12]. Therefore, a maximum distance of $5 \cdot d_B$ between the UFB has to be maintained within the cluster to preserve a sufficient supersaturation which prevents bubble shrinking. Applying the theory of Weijs et al. [Wei12] for UFB with a diameter of $d_B = 150$ nm, a minimum concentration in the magnitude of 10^{12} particles/mL is needed to stabilize the bubbles. A further approach is the stabilization of UFB by impurities within the system. Nanoparticles smaller than 12 nm do not get detected by the NTA and can be present in the used UPW. These particles would be adsorbed at the interface of the bubbles and create a strong mass transfer resistance. The so far achievable UFB concentrations result in gas hold-ups below $5 \cdot 10^{-5}\%$. Also, measurements of dissolved oxygen concentrations have shown no significant impact of the presence of oxygen UFB on the oxygen saturation concentration of the liquid phase [Kik09]. As long as the values of the gas hold-up cannot be increased drastically, an application of UFB to biocatalytic or chemical processes does not promise any advantage.

Since a direct proof of the gaseous state of the measured particles has not been provided so far, it is also possible that the NTA measurements detect clusters of solid nanoparticles.

The changes in concentration and diameter values observed during the stability analysis are effects which are also present for colloids and are described by the Ostwald ripening [Voo85, Shi15, Oh17]. Summarizing, the topic of bulk UFB is still a relatively new research field. Therefore, fundamental thermodynamical and physical investigations are required first, before applications to the field of process engineering can be considered. As a result, this work focuses in the following on microscopic bubbles and their impact on multiphase systems.

5.2 Evaluation and comparison of fine bubble aerated reactor concepts

To quantify the benefit of fine bubble aration for process engineering, different fine bubble generation techniques and reactor concepts are investigated in the following. To evaluate the impact of fine bubble aeration on the hydrodynamics and the mass transfer performance of bubble column, jet and stirred tank reactors, BSD and mass transfer coefficients are determined.

5.2.1 Hydrodynamic characterization of fine bubble aerated bubble column reactor

With regard to an application of fine bubble aeration to process industry, the bubble-bubble interaction in large bubble clusters is important. For the evaluation of swarm effects, a dense microbubble flow is characterized along the height of a laboratory scale bubble column reactor. To analyze phenomena such as bubble break-up and bubble coalescence, the bubble size distrubtion is measured at three different heights inside the bubble column. The investigations are done in oxygen saturated water so that mass transfer effects influencing the bubble diameters are neglected. The occurrence of break-up and coalescence changes the bubble size distribution having a direct impact on hydrodynamics an the mass transfer within the dispersed system. Therefore, it is a crucial aspect which has to be considered in the design process of a reactor system.

In figure 5.4, the characteristic mean diameters of the bubble swarm are plotted for the three measurement positions. For a comparison of bubble size distributions, no uniform parameter exists in the literature. Hence, the arithmetic mean diameter d_{50} as well as the Sauter diameter d_{32} are given. Taking into account the surface area and the volume of the dispersed phase, the Sauter diameter is often used in connection with mass transfer evaluations. A strong influence of the measurement position on both mean diameters is visible. With increasing reactor height the mean diameters are increasing as well, indicating coalescence of the bubbles. Between measurement port four and port three, there is nearly no difference in the Sauter diameter visible. The Sauter diameter is very sensitive to large bubbles so that strong deviations are caused by only a few extremely large bubbles captured within the measurements. Therefore, a deeper look at the bubble size distribution is recommended for a more precise evaluation of the bubble behavior.

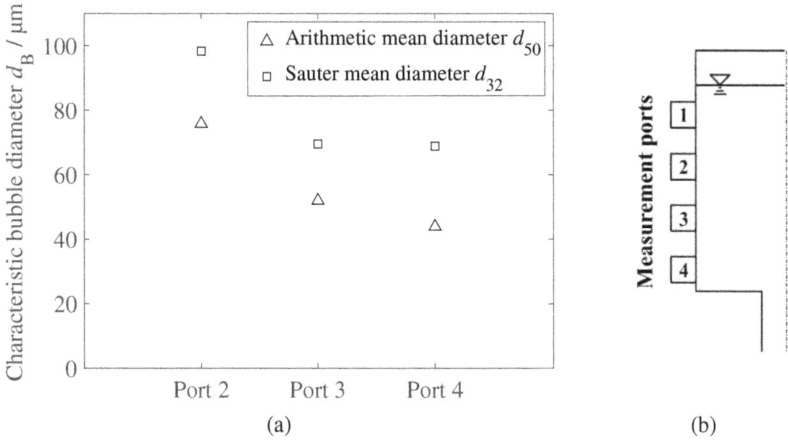

Fig. 5.4 (a) Evaluation of the characteristic mean diameters along the reactor height in a fine bubble aerated bubble column. (b) Illustration of the measurement positions

Figure 5.5 displays the bubble size distribution measured at all three reactor heights. For the lowest position (port four) at the entrance to the reactor, the bubble size distribution has a bimodal shape with a major peak at $d_B \approx 20$ µm and a minor peak at around $d_B \approx 40$ µm. Thus, a higher concentration of smaller bubbles occurs. A few bubbles larger than 100 µm are also present in the system which explains the relatively high Sauter diameter at measurement port four, compared to measurement port three. The bimodal shape of the BSD becomes less pronounced at measurement port three. Futhermore, a shift to larger bubbles caused by coalescence is visible. There are no longer bubbles smaller than 20 µm present, whereas the amount of bubbles larger than 40 µm has been increased. The major peak of the BSD is now located slightly below 40 µm. At the highest measurement position (port two) there is an ongoing shift towards larger bubbles visible. Furthermore, the weighting of the characteristic peaks of the bimodal distribution has shifted with the major peak now being located at the higher bubble diameter of around $d_B \approx 70$ µm.

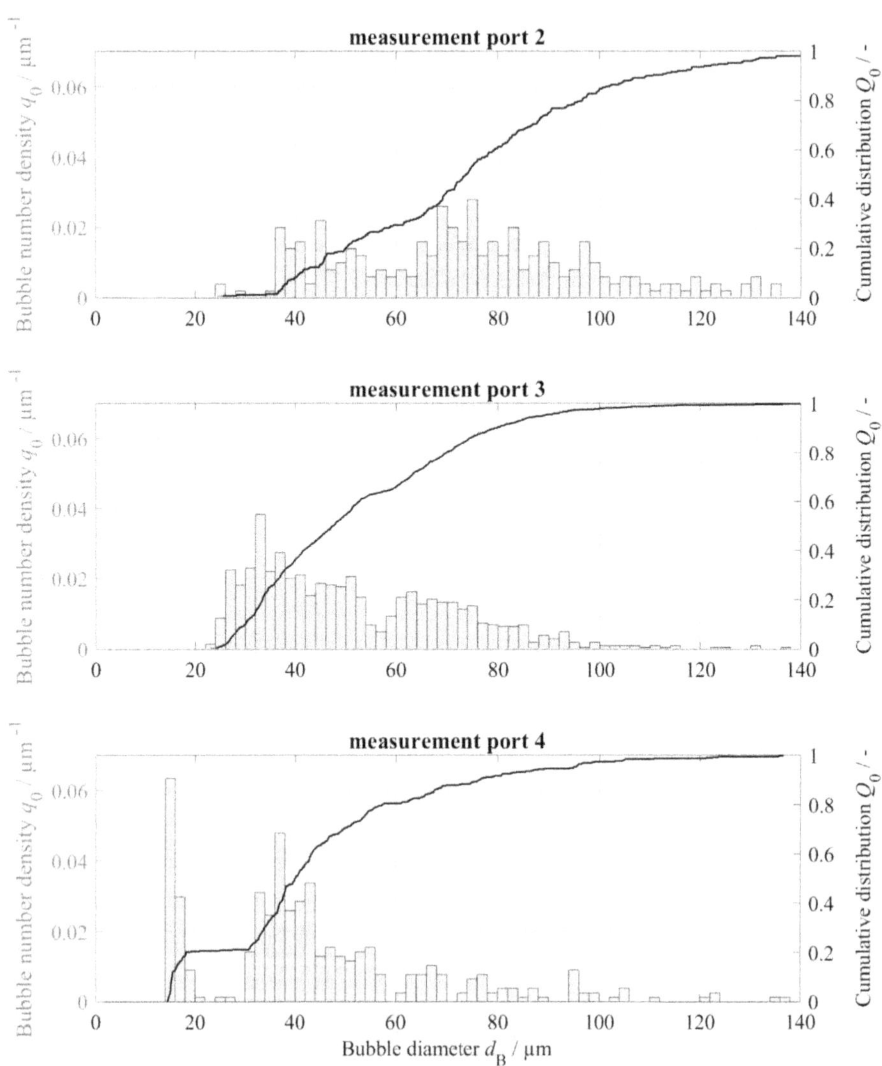

Fig. 5.5 Bubble number density and cumulative BSD at different heights of a fine bubble aerated bubble column reactor

The presented results are indicating a continuously shifting bubble size distribution which is dominated by coalescence effects along the height of a fine bubble aerated bubble column reactor. The observation of a distinct coalescence behavior of free rising microbubbles represents a significant gain in knowledge. Due to the form stability of those small scale spherical bubbles, coalescence has mostly been neglected so far in process design including microscale bubbles. The continuously changing BSD as shown in figure 5.4 causes inhomogeneities in the local gas hold-up due to different bubble rise velocities and has to be considered in the design of a process. Especially for pilot and industrial scale systems this can lead to the formation of compartments with different hydrodynamic characteristics.

5.2.2 Characterization of fine bubble aerated jet reactor systems

A large number of the commercially available fine bubble generators belong to the category of two-phase nozzles. Therefore, they are ideal for the application in jet reactor systems. In this chapter, two generators, one utilizing the swirl principle and another utilizing the ejector principle, are characterized with regard to achievable bubble size distributions and their mass transfer performance. For all experiments within this chapter, a deionized water/air system at a constant temperature of $T = 25$ °C is used. For the mass transfer measurements, the water is degased by aeration with nitrogen.

Evaluation of bubble size distribution and mass transfer coefficients

The BSD and mass transfer coefficients are measured for various operating parameters by changing the power input as well as the gas volume flow rate. For the characterization of a fine bubble generator, the produced bubble size and the width of the distribution are important parameters. Figure 5.6 shows a comparison of the BSD for both analyzed nozzles (swirl and ejector nozzle) for the same volumetric power input $P \cdot V^{-1} = 10$ kW/m^3 and a fixed gassing rate of $\dot{V}_{\mathrm{G}} = 100$ mL/min. The BSD for all investigated power inputs are shown in appendix B. Both BSD have the shape of a logarithmic normal distribution with the majority of the generated bubbles within the fine bubble regime below 100 µm. The BSD of the swirl nozzle is narrower than the one for the ejector nozzle with a significantly larger number of bubbles smaller than 100 µm. This is also reflected in the arithmetic mean diameter d_{50} as displayed in figure 5.7a. The mean diameter for the swirl nozzle is about 25% smaller than the one for the ejector nozzle. In addition, it can be seen that the volumetric power input, which is proportional to the liquid volume flow rate, has no influence on the generated bubble sizes for both fine bubble generators (compare also figure B.1 and B.2).

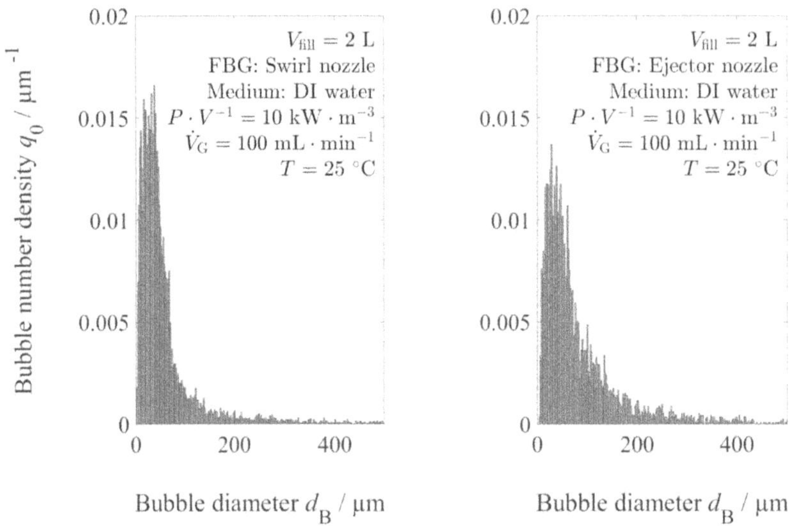

Fig. 5.6 BSD distribution for swirl nozzle (left) and ejector nozzle (right) in deionized water for $P \cdot V^{-1} = 10$ kW/m^3 and $\dot{V}_G = 100$ mL/min

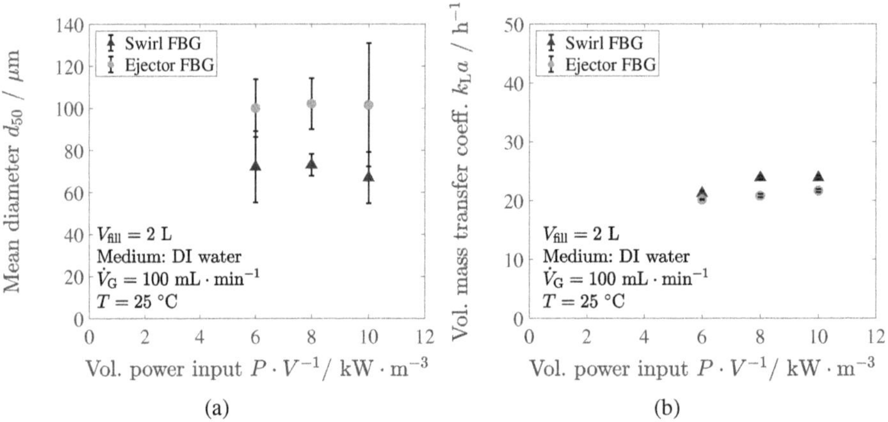

Fig. 5.7 (a) Influence of power input on mean bubble diameter for swirl and ejector nozzle. (b) Volume specific mass transfer coefficient for swirl and ejector nozzle in nitrogen saturated water as a function of the power input

The smaller bubble sizes produced by the swirl nozzle result in a larger surface area of the gaseous phase compared to the ejector nozzle which is beneficial for the mass transfer. Figure 5.7b shows the evaluation of the mass transfer performance by the measured volumetric mass transfer coefficients. As predicted from the bubble size distributions, the swirl nozzle achieves slightly higher values. Although the BSD remains constant for all investigated power inputs, an increase of the power input results in a small increase of the mass transfer coefficients as well. A higher power input comes along with a higher liquid volume flow rate causing higher local velocities inside the vessel. As a result, the convective mass transfer gets enhanced and the overall mass transfer performance becomes better. In addition, the mixing between the gas-liquid jet and the surrounding water inside the vessel also gets enhanced.

The mass transfer performance is controlled by the gas volume flow rate. The gas flow rate influences the gas hold-up of the system and correlates directly with the mass transfer coefficient. Experiments have shown that for both generators the volumetric mass transfer coefficient increases linearly with increasing gas volume flow rate.

5.2.3 Comparison of fine bubble aerated jet and stirred tank reactor concepts

To evaluate the efficiency of the fine bubble aerated jet reactor concept, a comparison to a stirred system with respect to their mass transfer performances is given in this section. The jet reactor setups investigated in section 5.2.2 operate at high volume specific power inputs of several kW/m^3 due to the required liguid flow rates for the injector nozzle FBG and the resulting pressure drop. To achieve equal volumetric power inputs with a stirred system, huge efforts have been made in the design of the STR. As mentioned in section 3.2.3 a three-stage configuration, as shown in figure 5.8b, consisting of two Rushton turbines and an up-pumping pitched blade is used. The pitched blade prevents the formation of a surface vortex at the high required stirrer frequencies. The Newton number of the system has been determined as $\mathrm{Ne} = 10.6$ and fine bubble aeration is implemented by a sintered frit made of 316L stainless steel with a mean pore size of 0.5 μm located below the lowest impeller. For both reactor concepts the liquid filling volume is equal with $V_{\mathrm{fill}} = 2$ L taking into account the volume of the hoses and the pump for the jet reactor system.

Evaluation of the bubble size distribution

To varify a present fine bubble aeration for the STR, the bubble size distribution is measured for the three investigated power inputs. The analyzed power inputs are at such a high level that there is no influence on the occuring bubble size within the system anymore. A detailed discussion on the bubble-impeller interaction and the acting mechanisms at fine bubbles will be given in section 5.3.1. As an example the BSD within the fine bubble aerated STR is given in figure 5.8a for a volumetric power input of $P \cdot V^{-1} = 10$ kW/m^3 and a gas volume flow rate of $\dot{V}_G = 100$ mL/min. The bubble number density is logarithmic normal distributed with the majority of the bubbles smaller than $d_B = 100$ µm and its peak at around $d_B = 70$ µm. Thus, the STR setup is classified as a fine bubble aerated system.

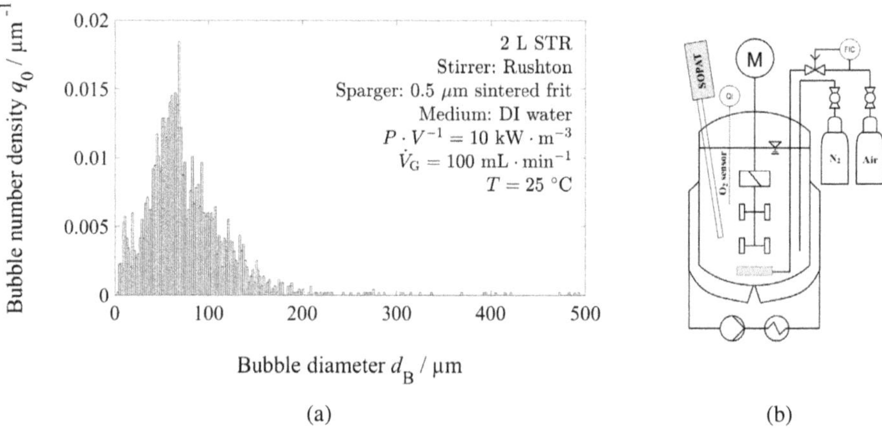

Fig. 5.8 (a) BSD distribution for membrane aerated STR in deionized water for $P \cdot V^{-1} = 10$ kW/m^3 and $\dot{V}_G = 100$ mL/min. (b) Scheme of the three-stage stirrer STR setup

Evaluation of mass transfer coefficients

Looking at the mass transfer performance of the STR, as displayed in figure 5.9, it can be seen that there is a significant influence of the power input on the mass transfer performance. The measured mass transfer coefficients for the STR setup reach twice the magnitude as those of the jet reactor setups at the same power input. For all reactor setups, the same gas volume flow rate has been used and just small differences in the bubble sizes are visible which cannot explain the huge difference in the mass transfer coefficients. Though, the flow field of the STR strongly differs from the one of the jet reactor, leading to an improved mass transfer. In a STR, flow structures are forming compartments in which the gas gets trapped. As a result, the residence time of the gas phase is enlarged, increasing the total gas hold-up of the system and with it the mass transfer coefficient. In these fine bubble aerated systems, the overall gas hold-up is below 1%. So, small changes can have already a big impact on the mass transfer coefficient. Also, a stirred system has a better mixing performance compared to the simple jet reactor setup used in this work. A fast mixing is beneficial in terms of balancing concentration gradients of the dissolved oxygen what maintains the maximum concentration gradient for the mass transfer from the bubbles into the liquid.

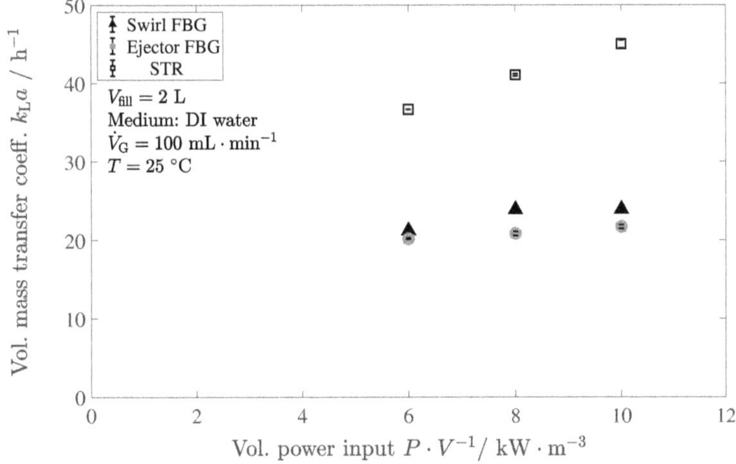

Fig. 5.9 Volume specific mass transfer coefficient for swirl nozzle, ejector nozzle and membrane aerated STR in nitrogen saturated water as a function of the power input

As a conclusion of evaluating different reactor concepts, the STR has shown to be the most promising system for the application of fine bubble aeration with regard to an efficient mass transfer. Using membrane spargers, fine bubbles are generated with a low energy consumption and only minor changes need to be made to the existing reactor system. The special features of the membrane aerated STR will be examined in greater detail in the following section.

5.3 Characterization of the membrane aerated stirred tank reactor concept

In section 5.2.3, the potential of the fine bubble aerated STR has been shown in contrast to other well-established fine bubble generators based on the principle of a two-phase nozzle. The formation of fine bubbles using porous membranes or sintered materials enables the integration of fine bubble aeration into the conventional designed STR setup by only changing the sparging device. This allows an easy application of those membrane aerated STR to existing processes. In this chapter the characteristics of the fine bubble aerated STR are analyzed in detail with regard to hydrodynamics and mass transfer performamces. The experiments focus on a single-stage system and point out the influence of impeller geometries, power input and gassing rate on the performance of the membrane aerated STR setup.

5.3.1 Hydrodynamics of membrane aerated stirred tank reactor

In an aerated STR the interaction between the dispersed gas phase and the continuous liquid phase is important for the characterization of such a two-phase system. The bubble size and its distribution (whether homogeneous or heterogeneous) is influencing the hydrodynamics of a system, in terms of the mixing behavior, as well as the overall mass transfer. Therefore, the determination of bubble size distributions and the identification of crucial parameters having an influence on it are of high importance.

The amount of gas that is supplied to the system is a central parameter for process controlling. In mass transfer limited processes for example, an increase in the available amount of oxygen within the system leads to a higher yield. Therefore, information on the performance of the sparging device for different gas volume flow rates are necessary. Figure 5.10 shows the BSD in a membrane aerated 2 L STR with a 0.5 μm sintered frit for gassing rates from 20 mL/min to 120 mL/min. A Rushton turbine is used for stirring and as liquid deionized water with Triton X-100 at its critical micelle concentration (CMC) is used. The sintered frit produces a narrow logarithmic normal bubble size distribution with its peak at around $d_B = 50$ μm for all gas flow rates. A change in the gassing rate does not lead to an influence on the BSD which remains constant. The bubbles which are formed by the sintered frit are too small so that flooding of the impeller does not occur and the system remains homogeneous for all investigated gassing rates.

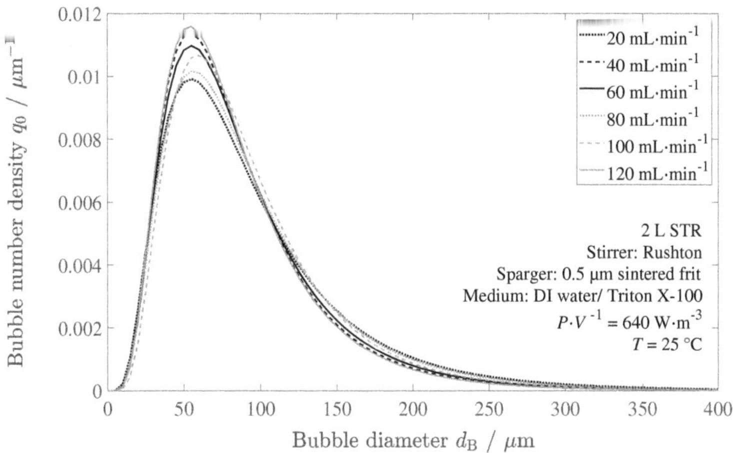

Fig. 5.10 Influence of gas volume flow rate on BSD for membrane aerated STR

At the same time the increasing gas flow rate results in a linear increase of the gas hold-up as shown in figure 5.11. Due to the constant BSD the specific interfacial area increases in the same manner what finally leads to an improved mass transfer rate according to equation 2.14.

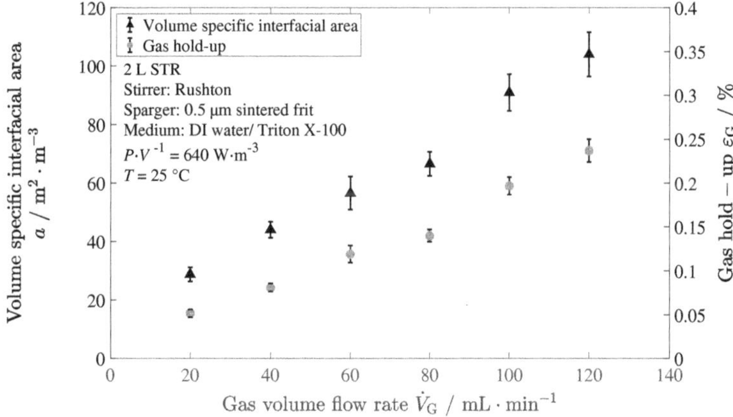

Fig. 5.11 Gas hold-up and specific interfacial area in fine bubble aerated system, obtained from endoscopic measurements

Beside the gassing rate, the power which is induced to the system is the second crucial parameter for controlling a process. In a STR the power that is induced by the impeller is proportional to the stirrer frequency as shown in equation 2.21. The influence of the power input on the BSD due to a dispersion of the gaseous phase is displayed in figure 5.12 for the fine bubble aerated system.

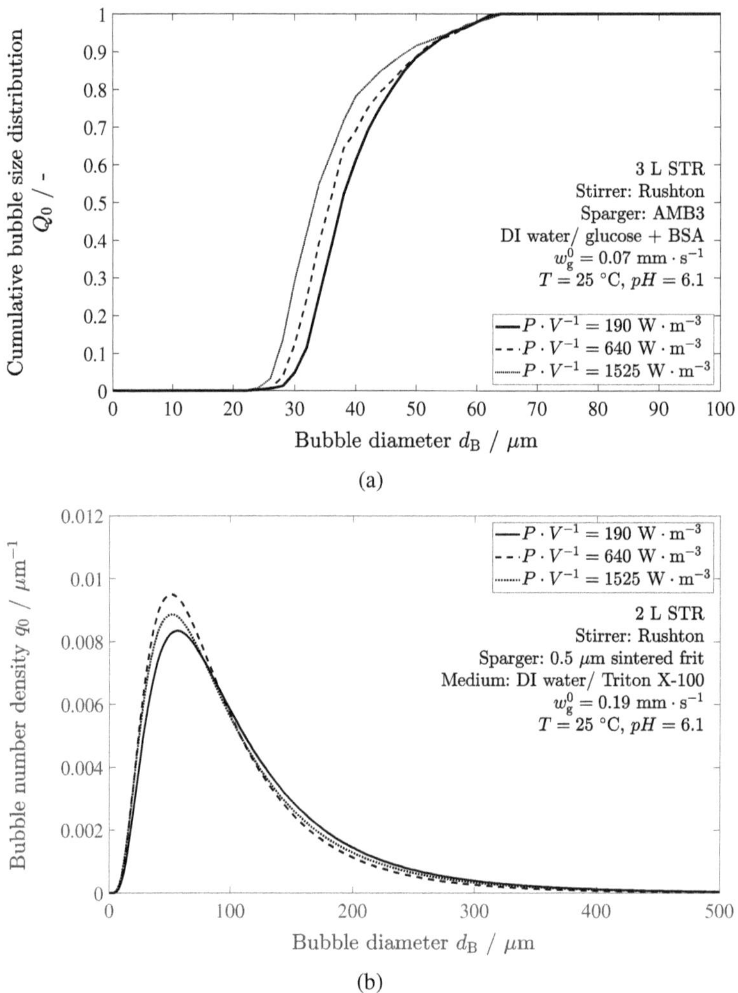

Fig. 5.12 Influence of power input on the BSD in fine bubble aerated STR with (b) and without (a) surfactant

Figure 5.12a shows the results obtained in deionized water. To study the interaction between the bubbles and the Rushton turbine, the AMB3 fine bubble generator which produces the highest microbubble number densities is used for aeration. The cumulative distribution displays that the minimum and maximum bubble diameter remain almost the same changing the stirrer speed. With increasing stirrer speed there is a slight decrease in diameter of the occuring bubbles within the BSD. A possible explanation for this change can be given regarding coalescence effects present in deionized water as explained in section 5.2.1. With higher stirrer frequencies the coalescence behavior of the system is reduced due to higher local shear stress induced by the Rushton turbine. Measurements in the Triton X-100 system support this conclusion. The surfactant which gets adsorbed at the interface of the bubbles reduces the coalescence behavior. Evaluations of the BSD for the Triton X-100 system at different power inputs are given in figure 5.12b for a membrane aerated STR setup. It can be seen that there is no significant influence of the power input on the occuring bubble sizes. The BSD are nearly congruent for the analyzed power inputs. These results are in good agreement to the work of Parthasarathy and Ahmed [Par94b] showing as well that intensive stirring does not lead to significant changes in bubble size for a system with non-coalescing conditions. As a conclusion, at the investigated power inputs the stirrer frequency does not show a significant influence on the bubble size distribution for fine bubble aerated STR. An explanation for this behavior is given regarding the Kolmogorov length scale

$$\eta_{\mathrm{K}} = \left(\frac{\nu^3}{\varepsilon}\right)^{\frac{1}{4}} \tag{5.1}$$

describing the size of the smallest eddies occuring in the turbulent system as a function of the kinematic viscosity ν of the liquid and the dissipation rate of turbulence kinetic energy ε in the system [Pop00]. The mean dissipation rate $\bar{\varepsilon}$

$$\bar{\varepsilon} = \frac{P}{V_{\mathrm{fill}} \cdot \rho_{\mathrm{L}}} \tag{5.2}$$

is calculated by the volumetric power input P/V_{fill} and the density of the system. According to Geisler et al. [Gei91], the maximum energy dissipation rate $\varepsilon_{\mathrm{max}}$ at the stirrer tips is 30 times higher than the average value in the vessel. For the investigated deionized water system, the Kolmogorov length scale is calculated using $\varepsilon_{\mathrm{max}}$ within the range of $\eta_{\mathrm{K}} = 19$ µm to $\eta_{\mathrm{K}} = 11$ µm for power inputs varying from $P \cdot V^{-1} = 190$ W/m^3 to $P \cdot V^{-1} = 1525$ W/m^3. With the Kolmogorov length in the same magnitude as the minimum bubble diameter it is concluded that the energy of the eddies is not sufficient to disperse the bubbles any further.

For a membrane aerated STR, two mechanisms have to be considered which influence the resulting BSD. On the one hand, there is the interaction between the bubbles and the induced shear forces in the area close to the impeller as described above. On the other hand, the impeller generates a liquid flow along the surface of the porous sparger which directly influences the bubble formation and the detachment of bubbles from the sparger. To quantify the impact of the liquid flow, the BSD directly above the 0.5 µm sintered frit is analyzed by high speed imaging. In figure 5.13, the change of the BSD is given for different spacings between the stirrer and the sintered frit. As reference, the BSD produced by the sintered frit without any liquid motion is given. For the reference case there is a broad logarithmic normal distribution of the bubbles with its peak at $d_B = 500$ µm. It can be seen that the liquid flow induced by the stirrer has a huge impact in reducing the bubble sizes. The liquid is shearing off the bubbles from the sintered frit during their formation. The additional applied force leads to a detachment of smaller bubbles. A smaller gap between the impeller and the sintered frit decreases the amount of larger bubbles generated by the sparger and creates a narrower BSD.

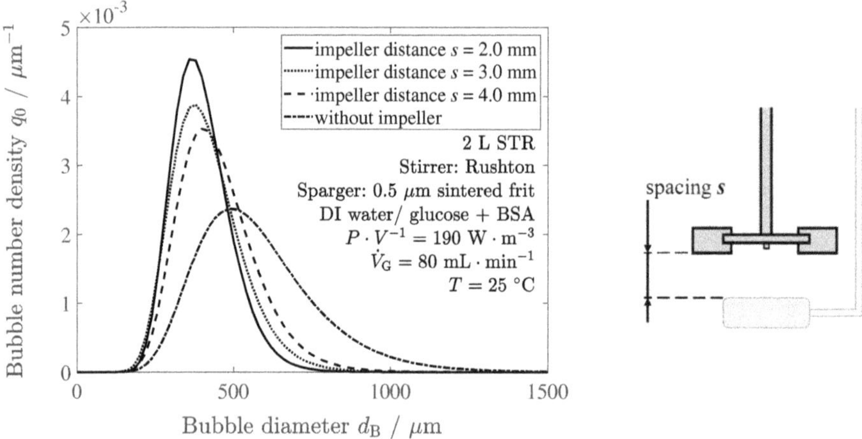

Fig. 5.13 Influence of the stirrer distance on the bubble formation in membrane aerated STR

Having shown that the liquid flow acting at the surface of the membrane sparger has a strong influence on the bubble formation, the behavior for different impeller geometries has to be analyzed. As described in section 2.3.1, impellers are divided, according to the generated flow pattern induced by stirring, in axially pumping and radially pumping impellers. To visualize the influence of the flow pattern on the bubble formation, the BSD in the Triton X-100 system is determined above the 0.5 µm sintered frit for a Rushton turbine (radially pumping) as well as a pitched blade and a segment impeller (axially pumping). Figure 5.14 displays the BSD for all three impeller types and three different power inputs. The sintered frit is located centrically between the bottom of the reactor and the impeller, so that the distance between the impeller and the surface of the sintered frit is fixed at 2.5 mm for all experiments.

At the lowest power input, the Rushton turbine creates the smallest bubble sizes of the three investigated impellers. With increasing power input, the BSD is moving towards smaller

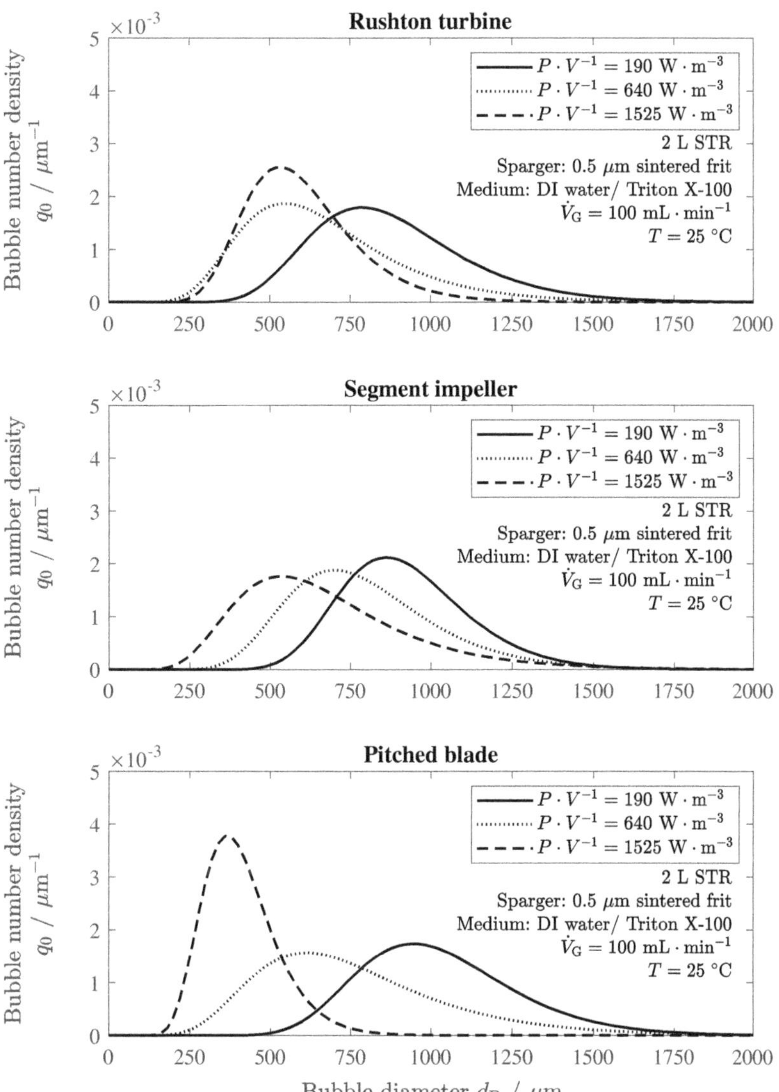

Fig. 5.14 Influence of impeller geometry on the bubble formation in membrane aerated STR

bubble sizes, but a stable state is reached quickly for the Rushton turbine. A further increase from $P \cdot V^{-1} = 640$ W/m^3 to $P \cdot V^{-1} = 1525$ W/m^3 does not lead to significantly smaller bubbles. The peak of the BSD stays constant and only the amount of larger bubbles is slightly reduced.

For the axially pumping impellers the increasing power input has a strong influence on the BSD within the total investigated range from $P \cdot V^{-1} = 190$ W/m^3 to $P \cdot V^{-1} = 1525$ W/m^3. In contrast to the Rushton turbine where the flow is redirected by the reactor wall, the axially pumping impellers directly force the liquid towards the sparger. Therefore higher stirrer frequencies have a stronger influence on the shear flow along the surface of the membrane spargers. For the segment impeller the increasing power input results in a parallel shift of the BSD towards smaller bubbles sizes. The main peak of the generated bubbles changes from 860 µm to 700 µm and finally reaches 520 µm what is similar to the occuring bubbles sizes for the Rushton turbine at the highest power input. The smallest bubbles are generated using a pitched blade impeller. At the highest power input, a sharp distribution of the bubble sizes gets created with its peak at $d_B = 360$ µm. For the Rushton turbine and the segment impeller, large bubbles at around 1 mm are present for all power inputs. Using a pitched blade impeller, it has been achieved that no bubbles larger than 800 µm get detached from the sintered frit.

As a conclusion, the initial bubble size in a membrane aerated STR is strongly dependent on the power input as well as the impeller geometry. For fine bubble aerated systems, axially pumping impellers are beneficial for achieving smaller initial bubble sizes in contrast to conventionally aerated systems where radially pumping impellers are needed for a better dispersion of the gaseous phase.

To determine the impact of the impeller geometry and the power input on the dispersed phase within the whole reactor, in addition to the measurements above the membrane, the BSD is determined at three different heights using the SOPAT Sc probe. As illustrated in figure 5.16, measurement position one is located directly beside the impeller. The other two measurement positions are equally spaced with $s = 50$ mm.

For the Rusthon turbine, the comparison of the measurements at the different locations (one, two and three) as well as the different power inputs (see figure 5.16) show that the mainly generated bubbe size is nearly constant for all measurements.

Looking at the BSD, it can be seen that the characteristic peak remains at around $d_B =$ 50 µm. Between position two and three there is no significant difference visible comparing all six BSD. Only at position one, directly beside the impeller, the power input is influencing the BSD. With increasing power input the BSD becomes narrower and converges against the one that is representative for the higher measurement positions. The observed change of

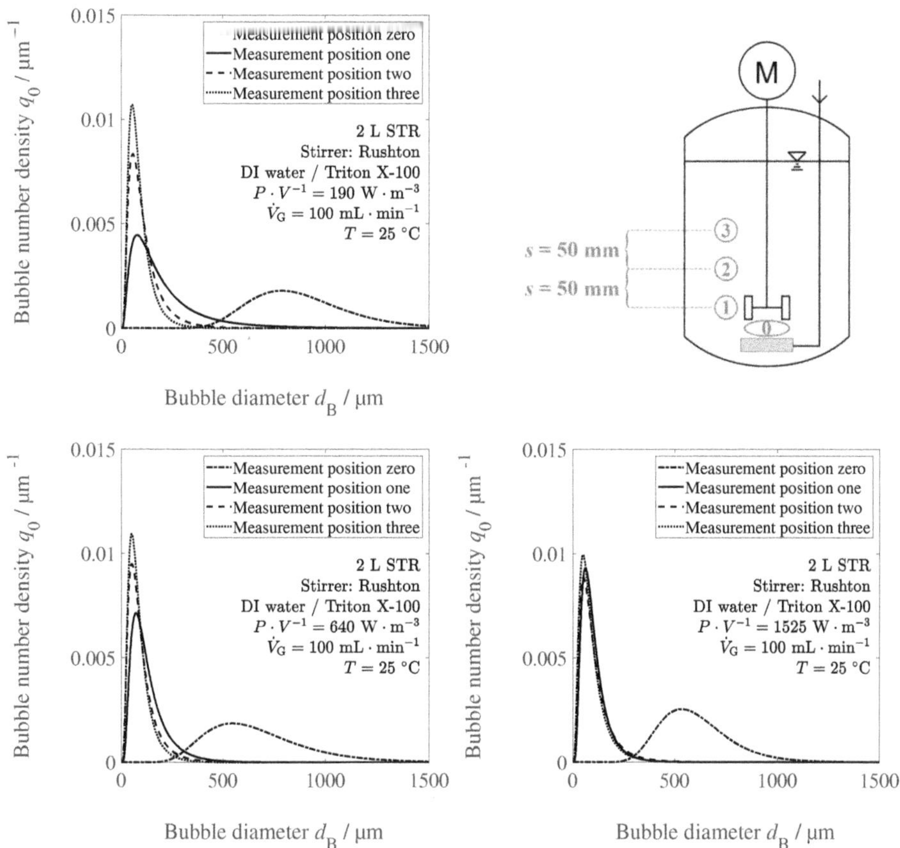

Fig. 5.16 BSD at four different locations of a STR and for three different power inputs. 2 L STR setup with a single-stage Rushton turbine

the BSD at position one results from the fact that not only the value of the maximum shear stress acting on the bubbles is influencing their dispersion but also the time which the bubbles are exposed to this shear stress. Due to the flow structure induced by the Rushton turbine, bubbles get trapped within the region of high energy dissipation rate close to the stirrer. At lower power input it takes longer until large bubbles are dispersed to their minimum size. Therefore, large bubbles are observed in the region close to the stirrer before they get divided into smaller bubbles and leave the scope of measurement position one. At higher power input, the break-up of bubbles is much faster and can already occure before entering the measurement area one. Due to the temporal resolution of the SOPAT system, the probability that larger bubbles get observed is reduced with increasing power input. These results prove the homogeneity of the gaseous phase for the fine bubble aerated STR as well as the negligible influence of the power input as previously described.

Comparing the BSD at the measurement positions one, two and three with the values obtained above the membrane sparger, a quantification of the two mechanisms acting on the bubbles during the aeration is given.

Axially pumping impellers distribute the liquid phase mainly axially and thus create a large-scale mixing within the vessel. As discussed in section 2.3.1, axially pumping impellers induce lower shear forces compared to radially pumping impellers and are therefore less suitable for dispersing of the gaseous phase. For fine bubbles, the initial bubble sizes are so small making the dispersion characteristics of the impeller less important.

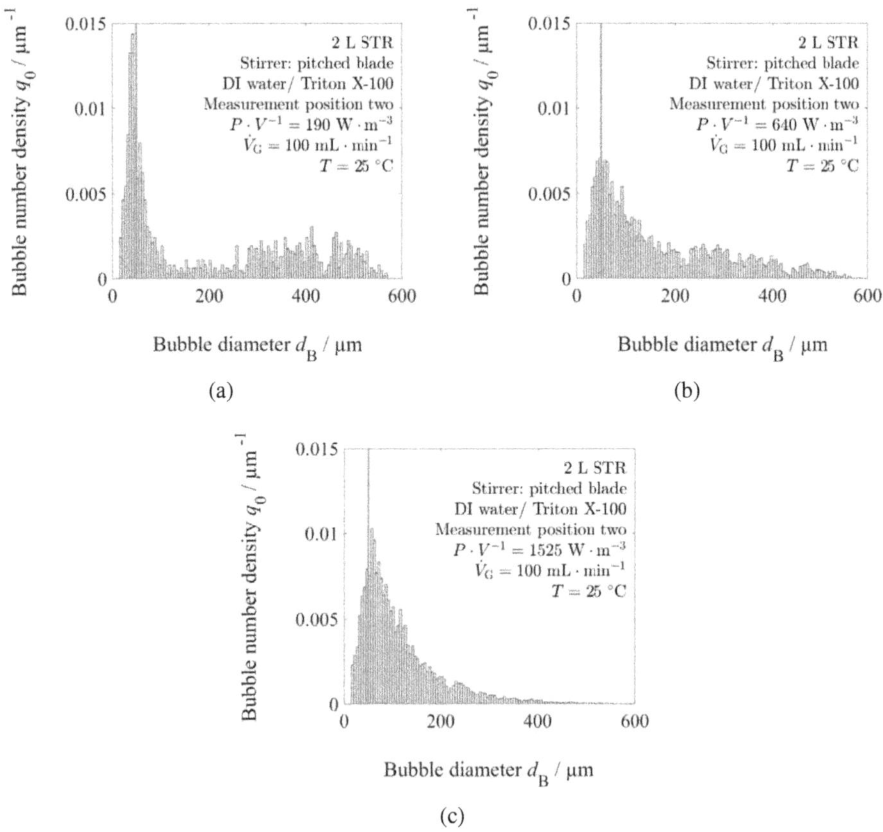

Fig. 5.17 BSD at measurement position two for the pitched blade impeller setup for $P \cdot V^{-1} =$ 190 W/m^3 (a), $P \cdot V^{-1} = 640$ W/m^3 (b) and $P \cdot V^{-1} = 1525$ W/m^3 (c)

Due to the good mixing in contrast to the Rushton turbine, the BSD for the axial pumping impellers remain constant along the reactor height for a fixed power input. The bubbles get distributed by the impeller over the whole reactor and do not get trapped close to the stirrer as described for the Rushton turbine. For the lowest investigated power input a bimodal BSD is observed within the reactor for both the pitched blade as well as the segment impeller. Figure 5.17 exemplarily shows the BSD at measurement point two for a pitched blade and all three analyzed power inputs. The distributions are representative for the segment impeller as well as for the measurement positions one and three. All measured BSD for the axial impellers are given in appendix C. With increasing power input the BSD becomes monomodal and logarithmic normal distributed. As pointed out for the Rushton turbine, the main characteristic peak of the distribution is constantly located at a bubble diameter of $d_\mathrm{B} \approx 50$ µm.

A comparison of all three analyzed impeller geometries shows only slight differences with regard to the achievable BSD having the same minimum bubble size. As shown in figure 5.18 for a volumetric power input of $P \cdot V^{-1} = 1525$ W/m^3, the measured BSD are nearly congruent and become independent from the impeller geometry.

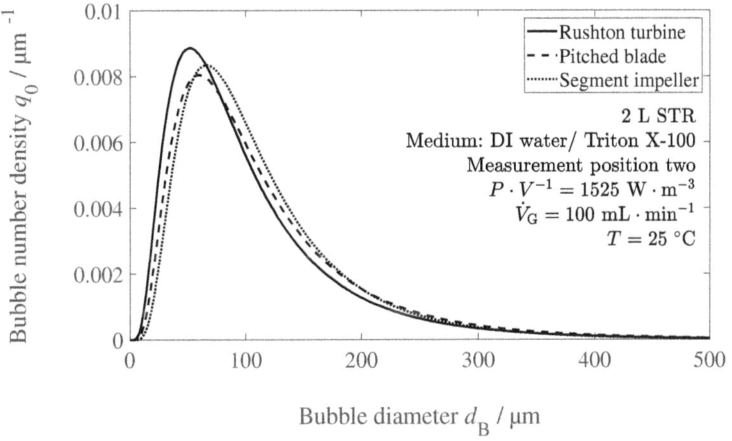

Fig. 5.18 Comparison of the BSD at measurement position three for all three analyzed impellers (Rushton turbine, pitched blade, segment impeller) at $P \cdot V^{-1} = 1525$ W/m^3

As a result for fine bubble aerated STR, the stirrer can be designed with a larger focus on the mixing performance or the prevention of high shear stresses as important for enzymatic processes. Good dispersing properties are not as necessary as for conventionally aerated systems due to the small initial bubble sizes.

5.3.2 Mass transfer behavior of membrane aerated stirred tank reactor

The aeration with microscopic bubbles offers large volume specific interfacial areas to achieve high mass transfer rates from the gaseous phase into the liquid. In this section, the mass transfer characteristics of microbubbles in stirred systems are analyzed. In addition to varying process parameters such as the gassing rate and the power input, different membrane spargers are also considered. The results presented in this section correspond in large parts to the published article of Matthes et al. [Mat20]. All mass transfer measurements within this section are done by initially stripping the system with nitrogen to reduce the dissolved oxygen concentration close to zero. The influence of the stripping gas on the mass transfer performance in a fine bubble aerated system is further investigated in the next chapter. In figure 5.19a, the volumetric mass transfer coefficient is shown for the 3 L STR setup equipped with a Rushton turbine and filled with 1.5 L of the Glucose/BSA system (compare section 3.2.3). For the microbubble aeration a SPG membrane with a mean pore size of 1 µm and a stainless steel sintered frit with a mean pore size of 2 µm are used. The oxygen $k_{\mathrm{L}}a$ is determined for various air flow rates. The $k_{\mathrm{L}}a$ for the two membrane spargers is reaching values close to $k_{\mathrm{L}}a = 90\ \mathrm{h}^{-1}$, showing the large potential of microbubble aeration. Having a reference for comparison, two measurement points using open tube aeration, which is often used in biocatalytic processes, and surface aeration are additionally performed as displayed in figure 5.19a. Comparing the data for the open tube aeration with the one for the sintered frit, the benefits of microbubble aeration become apparent. On the one hand, the sintered frit achieves the same mass transfer rate as the open tube with a diameter of 4 mm saving 60% of the gaseous phase. On the other hand, more than two times higher $k_{\mathrm{L}}a$ values are reached with the gas flow rate kept constant.

For both microspargers, the mass transfer rate increases with increasing gassing rate due to a growth in the total surface area of the gaseous phase. The sintered frit with 2 µm pores shows a better performance compared to the SPG membrane with 1 µm pores, having significantly higher volumetric mass transfer coefficients. This result appears unexpectedly on the first look, assuming a decrease in the pore size would result in smaller bubbles and higher $k_{\mathrm{L}}a$ values.

A higher $k_{\mathrm{L}}a$ can be caused by either a higher mass transfer coefficient k_{L} or a higher volume specific interfacial surface area a. Within the experiment it is only possible to measure directly the volumetric mass transfer coefficient $k_{\mathrm{L}}a$. The measurement of BSD and the information on the gas hold-up inside the measurement volume of the SOPAT probe (as explained in section 4.4) enable the possibility to get deeper information about the mass transfer behavior providing the data for a and k_{L} of the two-phase flow (compare table 5.1).

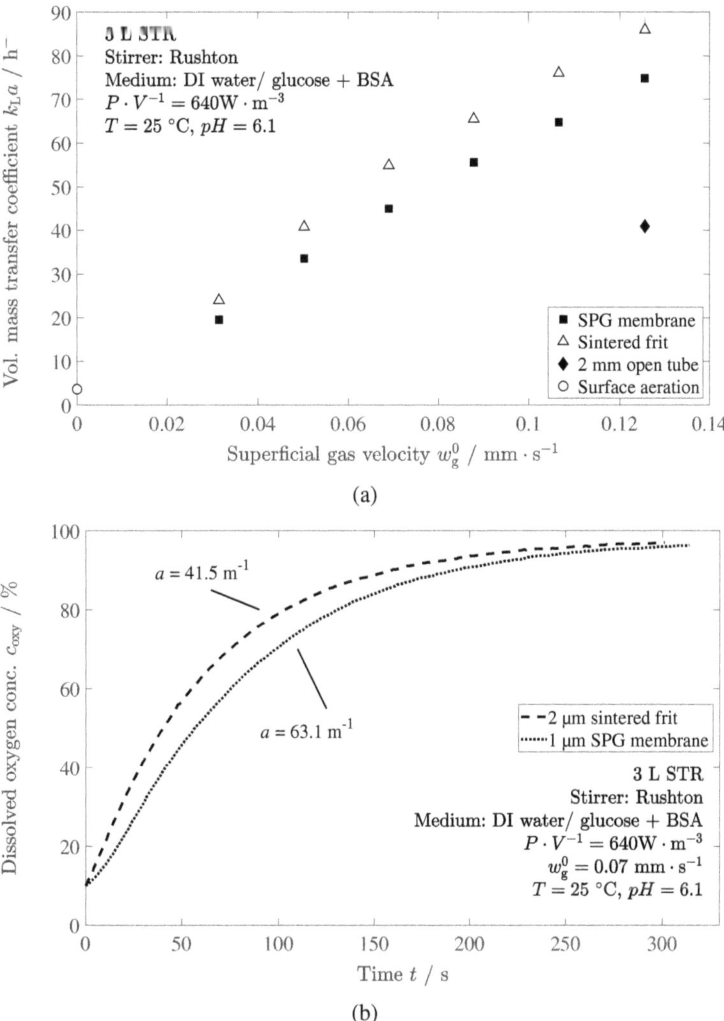

Fig. 5.19 (a) Volumetric oxygen mass transfer coefficients within the glucose BSA system for fine bubble and conventional aeration. (b) Dissolved oxygen concentration profile within the glucose BSA system for 1 μm SPG membrane and 2 μm sintered frit

The calculated data for a (as described in section 4.4) are in contradiction to the presented $k_\text{L}a$ values and cannot explain the difference of the two spargers. Also the gas hold-up of the SPG membrane system is larger compared to the sintered frit which achieves the higher $k_\text{L}a$ values. Since the total power input and the system properties remain constant for both spargers, an enhancement of the mass transfer could be explained by the surface deformation

Table 5.1 Volume specific interfacial surface area and gas hold-up for SPG membrane and sintered frit calculated from BSD for $w_g^0 = 0.07$ mm/s. Measured $k_L a$ values and resulting mass transfer coefficients k_L

	SPG membrane (1 μm pores)	sintered frit (2 μm pores)
Gas hold-up ε_G / -	0.0046	0.0033
Volume specific interfacial surface are a / m^{-1}	63.1	41.5
Volumetric mass transfer coefficient $k_L a$ / h^{-1}	45.0	54.9
Mass transfer coefficient k_L / $m \cdot h^{-1}$	0.71	1.32

of larger bubbles, resulting in a higher k_L value as described by Timmermann [Tim18]. It is assumed that the bubbles, which are not captured within the measurements, are still smaller than 1 mm in diameter having immobile surfaces (compare to [Cli78]). Therefore, an enhanced mass transfer due to surface deformation is not accurate for this case.

Considering the raw data for the $k_L a$, displaying the increase in the dissolved oxygen concentration over time during aeration, as shown in figure 5.19b, it can be seen that the mass transfer for the sintered frit is significantly faster than for the SPG membrane. Since both membrane spargers are reaching the same global saturation concentration during aeration, the reason for the measured $k_L a$ is found by looking into the BSD for both aerators. As described in the fundamentals of microbubbles (section 2.2.3), the partial pressure of the gas phase is not only influenced by the hydrodynamic pressure of the surrounding liquid. The Laplace pressure has an additional effect, influencing the saturation concentration at the gas-liquid interface, which cannot be neglected at these small bubble sizes. As shown in figure 2.8a, at a bubble diameter of around 15 μm the Laplace pressure reaches a characteristic value, resulting in an accelerated mass transfer rate. This local phenomenon is present in the microbubble aerated STR for all bubbles with diameters at this scale. Regarding the BSD for the two spargers as displayed in figure 5.20, the sintered frit produces a significant higher number of bubbles smaller than 15 μm than the SPG membrane. Therefore, the Laplace pressure inside the bubble is determined as the driving force for the mass transfer enhancement. This is also expressed through the mass transfer coefficient k_L, as shown in table 5.1, which is approximately 85% larger for the sintered frit system compared to the SPG membrane. It can be summarized for the microbubble aerated system, that for a high density of the smallest bubbles their mass transfer enhancing effect becomes dominant compared to the influence of the volume specific interfacial surface area.

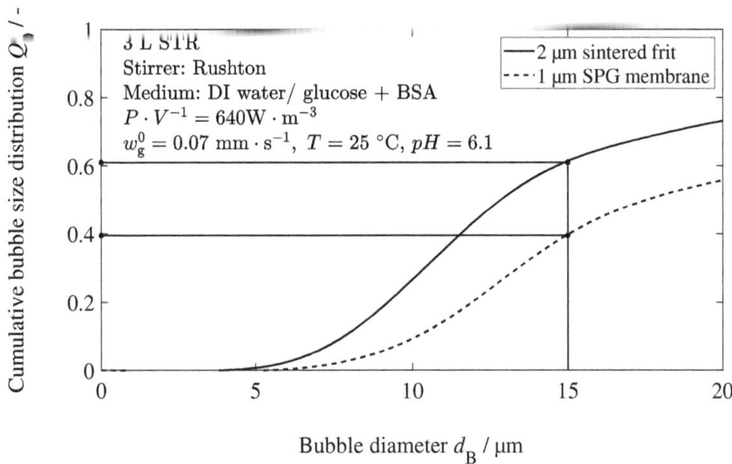

Fig. 5.20 Detailed view at the BSD for SPG membrane and sintered frit in the glucose BSA system

The volumetric power input has a relatively small influence on the mass transfer in a fine bubble aerated system. As shown in section 5.3.1, there are only slight changes in the BSD for different power inputs so that the volume specific interfacial area also remains nearly constant at the analyzed range for the power input. As discussed during the comparison of fine bubble aerated jet and stirred tank reactors (section 5.2.3), a higher power input is caused by higher stirrer frequencies influencing the flow structure as well as the mixing performance in the reactor. To quantify the influence of the power input on a single-stage Rushton stirrer, the volumetric mass transfer coefficient is measured within the Triton X-100 system for power inputs from $P \cdot V^{-1} = 40$ W/m^3 to $P \cdot V^{-1} = 640$ W/m^3. As shown in figure 5.21, only at small power inputs an increase of the power input leads to higher mass transfer coefficients. At around 200 W/m^3, the influence of the power input on the mass transfer becomes negligible. In contrast to the three-stage stirrer configuration where the flow structure can lead to an increase of the gas hold-up, in the system with a single Rushton turbine this effect does not occur.

The results on the mass transfer performance of a fine bubble aerated STR as presented in this section have pointed out the special mass transfer characteristics when microscopic bubbles are present. Due to the influence of the Laplace pressure, not only the volume specific interfacial are is of interest but also the number of the smallest bubbles has to be taken into account for evaluating the mass transfer performance. Furthermore, the power input cannot be used for controlling the mass transfer rate as for conventionally aerated systems due to a missing influence on the bubble size as figured out in section 5.3.1.

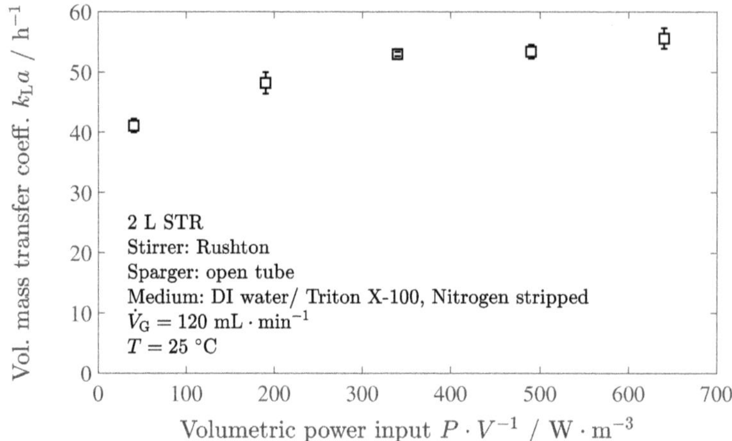

Fig. 5.21 Influence of the power input on the volumetric mass transfer coefficient in membrane aerated STR

5.4 Fundamental mass transfer characteristics of microscale bubbles

The previous presented results have shown the great potential of fine bubble aeration with regard to achieving high mass transfer rates or a more efficient utilization of the used gas. For a better understanding of the acting machanisms, the fundamental mass transfer characteristics of microscale bubbles are analyzed within this chapter.

Using laser-optical measurement techniques, the dissolved oxygen concentration fields in the vicinity of a single microscale oxygen bubble are investigated. Furthermore, the influence of counterdiffusion effects on the mass transfer is analyzed for single bubbles as well as bubble swarms, and models describing this influence are developed.

5.4.1 Visualization of local mass transfer effects at single microscale bubbles

The confocal laser scanning microscopy enables the quantitative analysis of dissolved oxygen concentrations in the vicinity of a microscale bubble with a spatial resolution of nearly one micrometer. Due to the limited temporal resolution of the CLSM, as discussed in section 4.3, the diffusive mass transfer from a fixed microscale oxygen bubble in nitrogen saturated deionized water is analyzed. In figure 5.22 the quantitative dissolved oxygen concentration fields around the shrinking bubble are displayed. The initial diameter of the oxygen bubble is approximately $d_B = 140$ μm and the mass transfer process is observed for 200 seconds.

The highest DO concentrations are reached directly at the gas-liquid interface. With increasing distance from the bubble, the DO concentration is strongly decreasing. In accordance to the film theory, the DO concentration at the interface corresponds to the oxygen saturation concentration $c^*_{O_2}$. Having a closer look at the DO concentrations close to the bubble it can be seen that the regions of high concentrations at the interface are disappearing after approximately 100 seconds. From this point on the mass transfer from the bubble into the liquid is completed and the diffusive mass transfer inside the liquid phase balances existing concentration gradients. Due to the low amount of transferred oxygen compared to the total liquid volume, there is no influence on the overall DO concentration visible after 200 seconds (comparing the concentration fields at $t = 0$ s and $t = 200$ s).

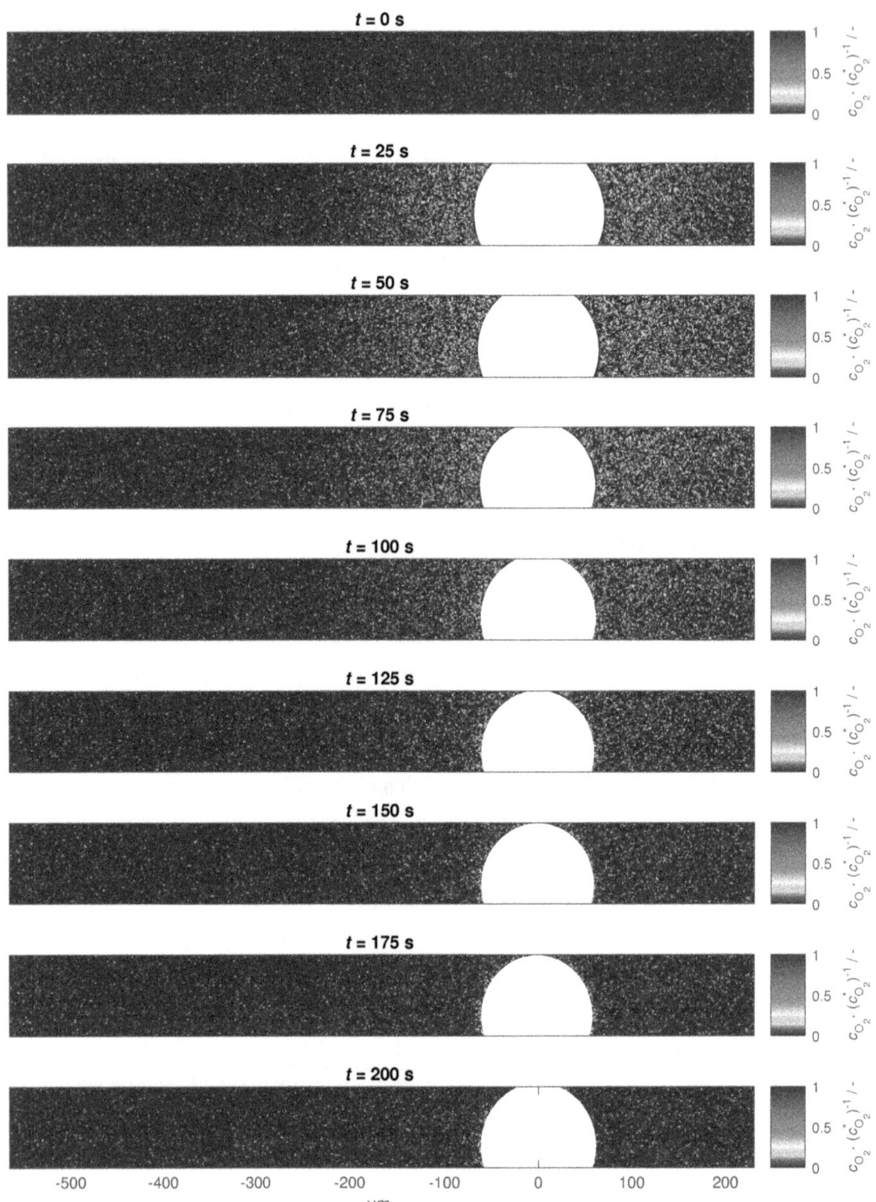

Fig. 5.22 Evaluated local oxygen concentration fields close to a fixed shrinking oxygen microbubble in nitrogen saturated water

Boundary layer thickness for diffusion

The data obtained from the CLSM measurements enable the determination of concentration boundary layer thicknesses for each measurement. Therefore, the DO concentration gradient between the gas-liquid interface and the bulk phase as displayed in figure 5.23 is analyzed.

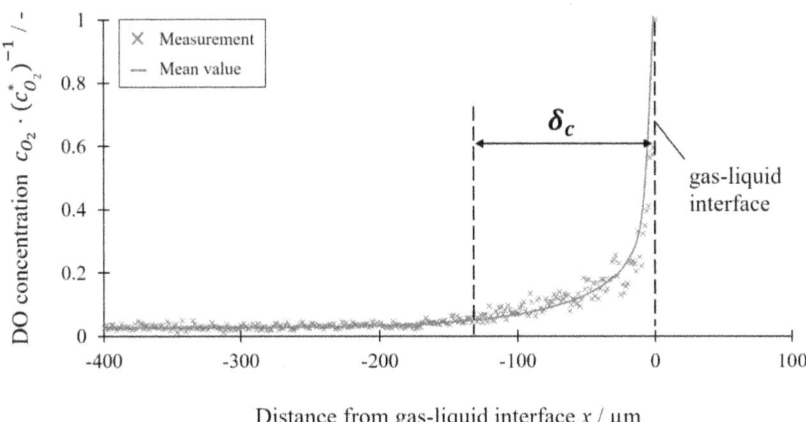

Fig. 5.23 Experimental data for the determination of the boundary layer thickness

The concentration boundary layer thickness is determined using the 95% criteria in analogy to the definition of the velocity boundary layer. As listed in table 5.2, the determination of the boundary layer thickness shows a good reproducibility for all six measurements. For the analyzed shrinking of a fixed microscale bubble, the concentration boundary layer is in the order of magnitude of $\delta_c = 125 \pm 20$ μm. Using equation 2.6, the mass transfer coefficient k_L is calculated for each experiment (see table 5.2).

A large uncertainty arises from the determination of the bubble size from the LIF pictures. The smooth transition of the gray values between the gaseous and the liquid phase complicates the edge detection to locate the interface. Nevertheless, the data allow a first estimation of the dimension of the boundary layer thickness δ_c and therefore of the mass transfer coefficient k_L for the diffusive mass transfer at a microscale bubble. Compared to the mass transfer coefficient in the fine bubble aerated STR as shown in table 5.1, the mass transfer coefficient for the fixed bubble is ten times smaller due to missing convective mass transfer. To visualize the influence of the liquid flow on the mass transfer, a rising microscale oxygen bubble is considered in the following.

Table 5.2 Approximate boundary layer thickness and mass transfer coefficient calculated from CLSM measurements

Measurement	Boundary layer thickness δ_c / µm	mass transfer coef. k_L / $m \cdot h^{-1}$
1	123.6	0.061
2	122.1	0.062
3	100.9	0.075
4	166.4	0.045
5	122.8	0.062
6	111.7	0.068

Boundary layer thickness for convection

Planar LIF allows the visualization of the dissolved oxygen transferred from a single microscale oxygen bubble during its rise in stagnant liquid. Figure 5.24 shows an oxygen bubble with a diameter of approximately $d_B \approx 398$ µm rising straight in nitrogen saturated DI water.

Due to the inhomogeneity of the illumination, a background correction is applied to reduce noise for more reliable information on the dissolved oxygen concentration. Typically for planar LIF measurements, there is a strong reflection of the LED light at the surface of the bubble complicating the evaluation directly at the gas-liquid interface. On the right side of the bubble, facing away from the light, the shadow is limiting the results. Using a bubble tracking algorithm, the rise velocity is determined from the measurements to $v_B = 43.5$ mm/s. Thus, the bubble Reynolds number is determined as Re $= 17.3$.

A comparison to the CLSM measurements at the fixed bubble shows the strong influence of the flow velocity at the interface on the mass transfer. The boundary layer thickness is strongly reduced compared to the case where only diffusion takes place ($\delta_{c,\text{fixed}}$), explaining the higher mass transfer coefficients. The transferred oxygen is visible within the thin cylindrical wake of the rising bubble. The observed structure of the wake is in good agreement to literature values [Was87]. The diameter of the wake is determined from the measurements as approximately $d \approx 83$ µm.

Due to these small dimensions, only qualitative results are obtained, having a state of the art light sheet with a minimum thickness of 1 mm. The illuminated areas behind and in front of the wake are outshining the areas quenched by oxygen. Therefore, the diameter of the concentration wake defines the maximum applicable light sheet thickness for measuring quantitative concentration fields. The results shown in figure 5.24 give an indication of the required optics. To observe the complete dissolution of a microbubble, the light sheet thickness has to be reduced down to a few micrometer as provided by light sheet microscopes used in the field of cell biology [Ola18, AS19].

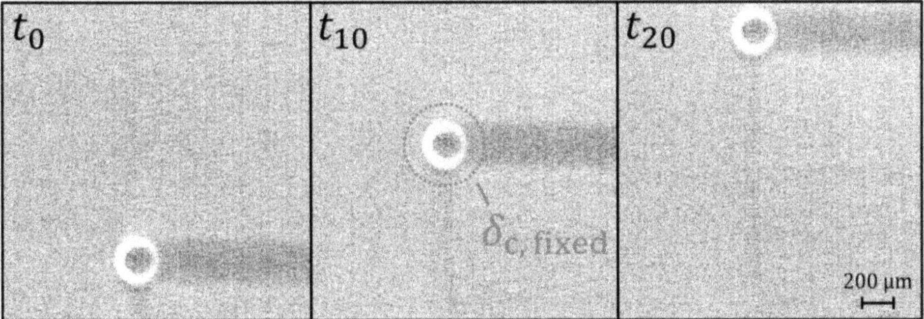

Fig. 5.24 Visualization of the dissolved oxygen concentration in the wake of a rising microscale bubble

Counterdiffusion effect at fixed microscale bubble

As previously discussed for the CLSM measurements of the fixed bubble, the duration of the mass transfer process is observed considering the DO concentration fields presented in figure 5.22. When observing the bubble shrinking during the measurement, the end of the mass transfer process is determined as well. Therefore, the bubble diameter d_B is tracked over time. As shown in figure 5.25, the bubble diameter reaches a constant value 80 to 100 seconds after the bubble formation being in good agreement to the conclusion from the analyzed concentration fields.

The fact that the oxygen bubble is not dissolving completely within the water is caused by simultaneous counterdiffusion of the dissolved nitrogen from the water into the bubble. At the end of the mass transfer process, all oxygen within the bubble is transferred and a pure nitrogen bubble which is in equilibrium with the surrounding liquid remains. This result fits well the model of Worden and Bredwell [Wor98] for the shrinkage of a microbubble where a non-transferred gas is present as described in section 2.2.3. The CLSM measurements point out an impact of counterdiffusion effects on the mass transfer performance of a single microscale bubbe. The influence of counterdiffusion on mass transfer is further analyzed and mass transfer models are developed for single bubbles as well as fine bubble swarms.

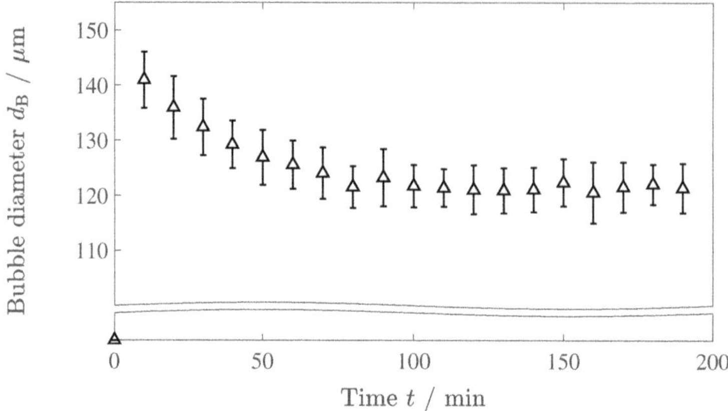

Fig. 5.25 Shrinking rate of fixed oxygen microbubble observed during CLSM measurements

5.4.2 Modeling of counterdiffusion effects at a single microscale bubble

The above described diffusive mass transfer of the dissolved oxygen within the liquid phase gets illustrated analyzing the DO concentration at a certain point in the liquid during the whole measurement. In figure 5.27, the DO concentration close to the interface at a fixed distance of $x = 70$ µm from the center of the bubble is plotted for the data presented in figure 5.22 with an initial bubble diameter of $d_B = 136$ µm. The measured DO concentration is a response to the oxygen which is initially supplied by the single microscale bubble. Around 15 seconds after the bubble formation, the DO concentration shows a strong increase and reaches its maximum value after additional 15 seconds. Afterwards there is a continuous decrease back to the initial DO concentration in the liquid.

The concentration profile is described using the analytical solution of the 3D diffusion equation as given by equation 2.7. In contrast to the analytical solution, for a dissolving bubble the initial oxygen supply cannot be described by a single Dirac delta function. In the case of a shrinking bubble, oxygen is continuously supplied until the mass transfer is completed. Due to the finite volume of the bubble and the acting counterdiffusion, the oxygen concentration within the bubble is decreasing over time. To consider the impact of the counterdiffusion within equation 2.7, a model for the initial concentration c_0 is developed. Therefore, the change of the DO concentration at a specific position R in the liquid $c(R,t)$ is described at each time by the superposition of its value for time-shifted Dirac delta functions with decreasing concentration $c_0(t)$ as an initial pulse. In figure 5.26 a quantitative illustration describing the developed model is given as an example. The different initial concentrations

(5.26a) and the resulting concentration responses which overlap each other (5.26b) are schematically shown.

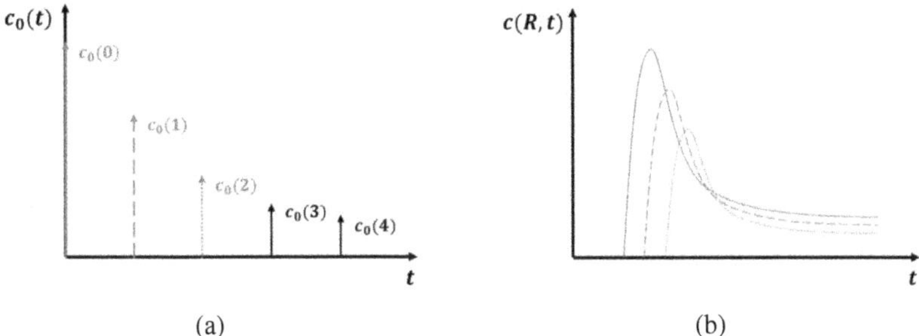

Fig. 5.26 Sketch for the explanation of superimposed Dirac delta functions (a) and the resulting concentration responses (b)

In figure 5.27, the modified solution of the 3D diffusion equation, as described above, shows an excellent fit to the measured concentration profile. This supports the assumption that the oxygen mass transfer from the fixed bubble into the liquid is driven by diffusion, and convection due to liquid being displaced during the bubble formation is negligible. The continuously decreasing initial oxygen concentration is modeled using an exponential approach. The parameters *a*, *b* and *c* are determined experimentally (see figure 5.27).

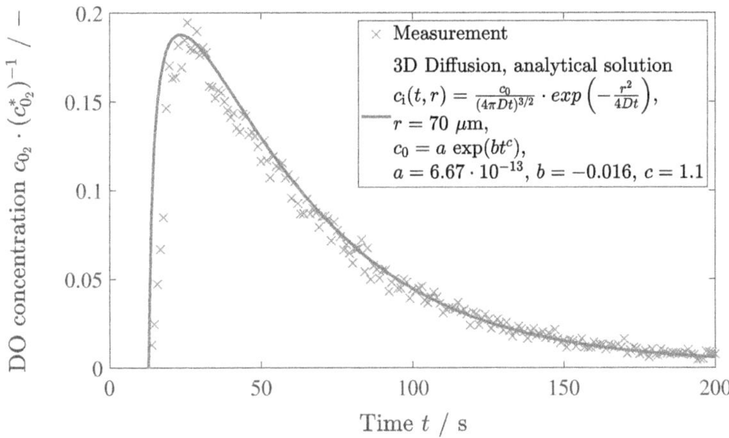

Fig. 5.27 Comparison of the measured local oxygen concentration with an analytic solution of the 3D diffusion equation including a model considering the counterdiffusion of nitrogen

5.4.3 Evaluation of counterdiffusion effects in fine bubble swarms

As pointed out so far, the mass transfer mechanisms for small scale bubbles differ from conventional theories. Due to the acting surface tension, at microscopic scale the Laplace pressure influences the mass transfer. Therefore, a changing bubble diameter has an impact on the mass transfer rate. As pointed out by Worden and Bredwell [Wor98] in a gas mixture, the amount of an insoluble component that influences the partial pressure of the oxygen and the shrinking rate of the bubble has to be taken into account (compare figure 2.9). Iwakiri et al. [Iwa17] have visualized first this influence by comparing the shrinking rate of a single oxygen microbubble in vacuum treated water and nitrogen stripped water. Having a high concentration of dissolved nitrogen in the water, not only oxygen from the bubble is transferred into the liquid but at the same time nitrogen diffuses into the bubble, reducing the shrinking rate of the bubble as previously discussed in section 5.4.2.

For a deeper investigation of this counterdiffusion effect on fine bubble swarms, different degassing methods are analyzed in this section and the influence on the mass transfer for different bubble sizes is quantified. All experiments are executed in the 2 L STR setup with a single Rushton turbine and the aqueous Triton X-100 solution. For fine bubble aeration, the 0.5 μm sintered frit is used, and conventional aeration with macroscopic and shape oscillating bubbles is realized using a 4 mm open tube sparger. To remove the dissolved oxygen from the water the system is stripped with nitrogen, argon or helium. All gases have a different solubility in water as given in table 5.3. Therefore, after the stripping process the concentration of the stripping gas which gets dissolved in the water has a different level depending on its solubility. As a reference, the physically degassed system, where no dissolved gas is present and hence no counterdiffusion takes place, is used.

Table 5.3 Solubility of the stripping gases and oxygen in water at $T = 25$ °C

Gas	Solubility S / $\mathrm{mg \cdot L^{-1}}$
Helium	1.5
Nitrogen	17.5
Argon	53.0
Oxygen	8.2

For a fixed power input of $P \cdot V^{-1} = 640$ W/m^3, the fine bubble aerated system reaches for all investigated degassing methods and all air gassing rates higher volumetric mass transfer coefficients than the conventionally aerated system. As illustrated in figure 5.28, for both spargers, the physically degassed system achieves the highest $k_{L}a$ values followed by the nitrogen stripped system and the argon stripped system showing the lowest mass transfer coefficients. The influence of the counterdiffusion is more visible for the fine bubble aerated system. For the open tube aeration, the argon stripped system reaches nearly the same $k_{L}a$ values achieved by the nitrogen stripped system. Due to the influence of the counterdiffusion on the partial pressure of the oxygen within the bubble, the initial bubble size in the system is important for the impact on the mass transfer. For larger bubbles the counterdiffusion has less influence on the mass transfer rate due to a smaller impact of the Laplace pressure on the total pressure inside the bubble. Additionally, the amount of gas transferred into the bubble is negligibly small to the total volume of the bubble.

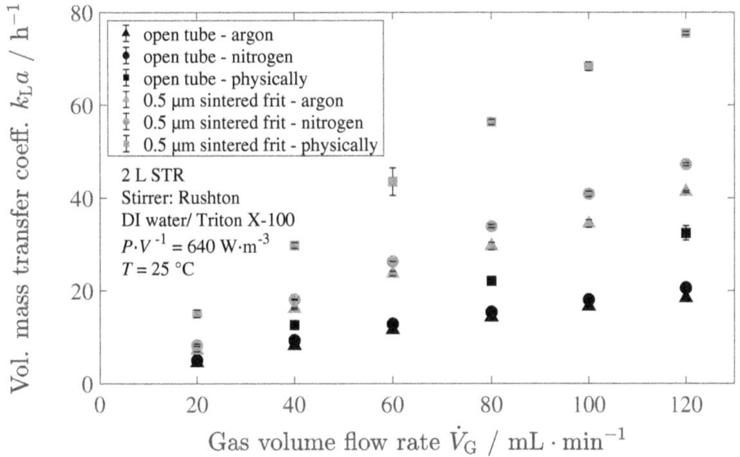

Fig. 5.28 Influence of the degassing method on the volumetric oxygen mass transfer coefficient for fine bubble and conventionally aerated STR

Comparing the BSD for both spargers from figure 5.29, it can be seen that the open tube creates 12 times larger bubbles compared to the sintered frit. The bubble sizes are logarithmic normal distributed for both spargers. The modal value for the BSD of the sintered frit is given by $d_{B} = 50$ µm, while for the open tube its value is at around $d_{B} = 500$ µm. The larger bubbles in the system using the open tube sparger point out the importance of the bubble size on the impact of the counterdiffusion effect. But, still for the larger bubble sizes in the conventionally aerated system, the mass transfer coefficients measured by the most frequently applied gas stripping method are nearly 50% smaller than the reference values obtained from the physically degassed system.

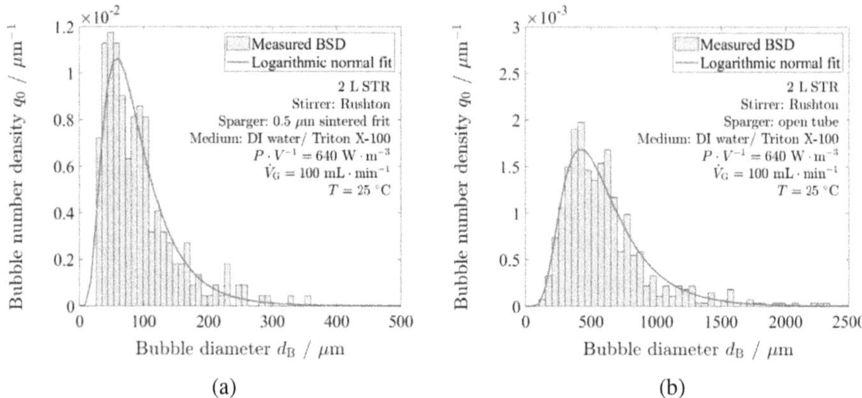

Fig. 5.29 Comparison of the BSD for sintered frit (a) and open tube (b) in the Triton X-100 system for $P \cdot V^{-1} = 640$ W/m^3 and $\dot{V}_G = 100$ mL/min

For a better quantification of the influence of bubble size and shape on the counterdiffusion effect, the conventionally aerated STR setup is used. The mass transfer coefficient is determined for all stripping gases (helium, nitrogen and argon) and the physically dagassing method for different volumetric power inputs reaching from $P \cdot V^{-1} = 40$ W/m^3 to $P \cdot V^{-1} =$ 640 W/m^3. As displayed in figure 5.30, for the open tube the bubble size and shape strongly depends on the power input. For a small power input of $P \cdot V^{-1} = 40$ W/m^3, shape oscillating bubbles of a few millimeter are generated. With increasing power input the bubbles become smaller and more spherical.

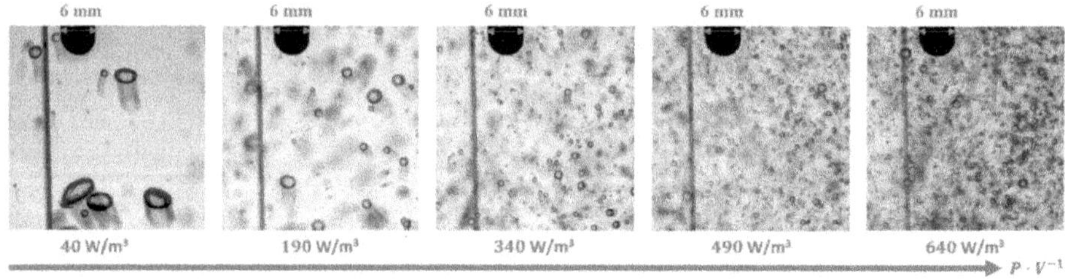

Fig. 5.30 Snapshots of bubble sizes in conventionally aerated STR in dependence of the power input

The changing mean diameters of the bubbles within the conventionally aerated stirred tank reactor are given in table 5.4 for all investigated power inputs.

Table 5.4 Arithmetic mean diameter and Sauter diameter for the open tube aerated system for various power inputs

Vol. power input $P \cdot V^{-1}$ / $W \cdot m^{-3}$	Arithmetic mean diameter d_{50} / µm	Sauter diameter d_{32} / µm
40	930	5005
190	751	2244
340	580	1028
490	437	786
640	394	484

Figure 5.31 shows the results for the mass transfer mesurements in the open tube setup for all power inputs and degassing methods. In general, the volume specific mass transfer coefficient increases with increasing power input due to the above described reduction in the bubble size and the resulting increase in the volume specific interfacial area.

For the lowest power input where the bubbles appear shape oscillating, the measured $k_L a$ values are equal for all degassing methods. Due to the larger bubble volume, the amount of the stripping gas that is transferred into the bubble does not lead to a significant change of the total bubble volume and its diameter. Also, the changing Laplace pressure has no significant influence on the total pressure inside the bubble for these large dimensions. Therefore, the partial pressure of the oxygen remains nearly constant. Furthermore, for large oscillating bubbles the deformation of the bubble surface is the dominating force driving the mass transfer as described by [Tim18] and not the Laplace pressure.

Without occurring bubble deformations and with decreasing bubble size, the influence of the stripping method gets visible. For a mean Sauter diameter smaller than $d_B \approx 1000$ µm there are significant differences in the measured $k_L a$ values due to the counteriffusion of the stripping gas. The highest $k_L a$ values are reached for the physically degassed system, and with increasing solubility of the stripping gas the $k_L a$ values are decreasing. The higher the solubility of the stripping gas, the more of it gets dissolved during the stripping process. As a result, there is a strong concentration gradient for the stripping gas from the water towards the bubble that correlates with the amount of gas diffusing into the bubble.

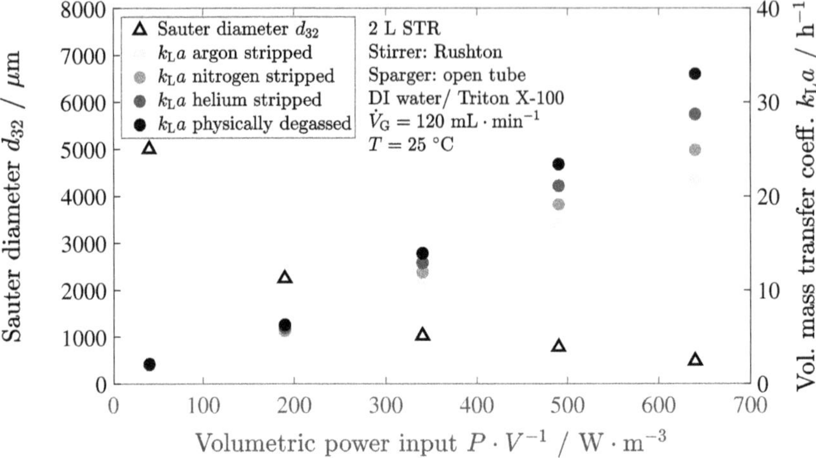

Fig. 5.31 Volume specific mass transfer coefficients and Sauter diameters for conventionally aerated STR as a function of the power input and the degassing method

For fine bubble aeration, simultaneous measurements of the BSD and the volume specific mass transfer coefficient are done visualizing the amount of the stripping gas which is transferred into the bubble. As shown in figure 5.32, for the fine bubble aerated system the influence of the counterdiffusion on the k_La values is visible for the smallest as well as for the largest power input due to the small bubble sizes.

Comparing the measured k_La values with the Sauter mean diameters that are measured during aeration, it can be seen that the physically degassed system has the lowest Sauter diameter. In table 5.5, the values are given for a power input of $P \cdot V^{-1} = 40$ W/m^3. With increasing solubility of the stripping gas, the Sauter diameter of the dispersed gas phase increases as well, caused by the counterdiffusion. The more gas is diffused into the bubble, the lower the partial pressure of the oxygen and therefore the oxygen saturation concentration at the interface get. Comparing the oxygen concentration profiles during the aeration, there is no acceleration or deceleration visible for neither of the analyzed degassing methods. This indicates that the counterdiffusion process is completed much faster compared to the oxygen mass transfer. So the different mass transfer performances are equal to the one which are observed for different initial gas mixtures.

The results of the counterdiffusion experiments presented in this section point out how important it is to capture all components that are present in the system. For fermentation processes for example, gases like carbon dioxide are generated during the process which influences the overall mass transfer. The information on the counterdiffusion can help to modify mass transfer measurements during the design of a process so that the real process is better predictable and can thus be reproduced more accurately.

Table 5.5 Sauter diameter and volume specific mass transfer coefficient for different degassing methods for fine bubble aerated system at $P \cdot V^{-1} = 40$ W/m^3

Degassing method	Volumetric mass transfer coefficient $k_L a$ / h^{-1}	Sauter diameter d_{32} / µm
Physically degassed	53.85	271
Helium stripped	40.30	279
Nitrogen stripped	35.51	286
Argon stripped	29.67	310

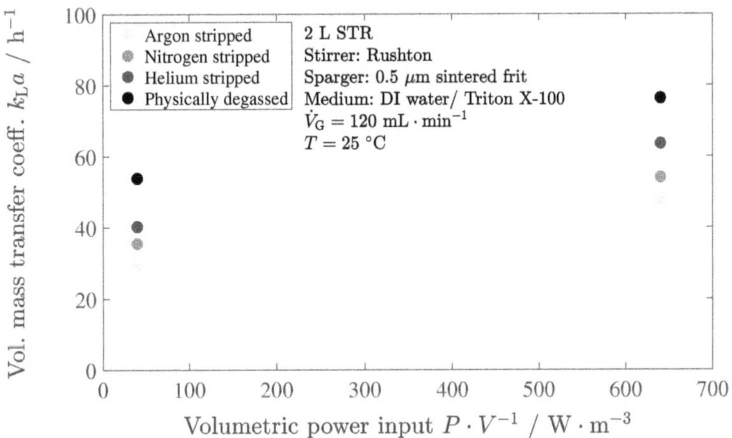

Fig. 5.32 Volume specific mass transfer coefficients for membrane aerated STR as a function of the power input and the degassing method

5.4.4 Modeling of counterdiffusion effects in fine bubble swarms

As discussed in section 5.4.3, the degassing method has a strong influence on the measurements of volume specific mass transfer coefficients especially for small spherical bubbles. Due to the counterdiffusion of the stripping gas, the volume of the bubble increases, leading to a change in the composition of the gas mixture inside the bubble. The oxygen gets diluted and its partial pressure decreases. As pointed out in section 5.4.3, in a bubble swarm with $d_{32} \leq 1000$ µm, the amount of stripping gas diffusing into the bubble is large enough compared to its initial volume to significantly change the partial pressure of the oxygen. For microscopic bubbles additionally the influence of the Laplace pressure on the partial pressure of the oxygen has to be taken into account, which is decreasing with increasing bubble size. As a result, in a fine bubble aerated system for each bubble size, a specific saturation concentration has to be defined for a correct description of the mass transfer according to the film theory. Experimentally, only global dissolved oxygen concentrations are measured and the global saturation concentration of the water/air system is determined. Due to Henry's law, the global saturation concentration remains constant and the local deviations in the saturation concentration at the interfaces of the bubbles are not taken into account for the calculations of the mass transfer coefficients. Thus, the oxygen diffusion coefficient gets apparently reduced, resulting in smaller k_L values calculated from the measurements.

In table 5.6, the k_L values calculated form the mass transfer measurements in the fine bubble aerated Triton X-100 system (results given in figure 5.32) are shown. For the determination of the corresponding volume specific interfacial area, the gas hold-up has been determined using the SOPAT data for the BSD as explained in section 4.4.

Table 5.6 Mass transfer coefficient for different degassing methods obtained in fine bubble aerated STR system

Degassing method	Mass transfer coefficient k_L / $m \cdot h^{-1}$	Measurement deviation Δk_L / $m \cdot h^{-1}$
Physically degassed	1.05	±0.079
Helium stripped	0.90	±0.054
Nitrogen stripped	0.74	±0.101
Argon stripped	0.67	±0.094

To include the influence of the stripping gas on the obtained mass transfer coefficients, a model is developed using dimensionless analysis which includes the dependence between the solubility of the stripping gas and the overall calculated mass transfer coefficient. As a reference, the physically degassed system ($k_L = k_{L,0}$) is used. Figure 5.33 dispalys the defined dependency which is described by a fractional-rational function of the form

$$\frac{k_L}{k_{L,0}} = \frac{a \cdot D_{O_2} \cdot S \cdot \nu_L^{-1} + b \cdot \rho_L}{S + b \cdot \rho} \tag{5.3}$$

with the substance data D_{O_2}, ρ_L, ν_L and the solubility of the stripping gas as variable parameter. For the analyzed fine bubble aerated 2 L STR setup with an aqueous solution of the standard surfactant Triton X-100 at its critical micelle concentration, the parameters a and b are defined as $a = 2.7 \cdot 10^2$ and $b = 2.7 \cdot 10^{-6}$.

The developed correlation enables the conversion of the mass transfer coefficient between the different stripping methods, characterized by the solubility of the stripping gas.

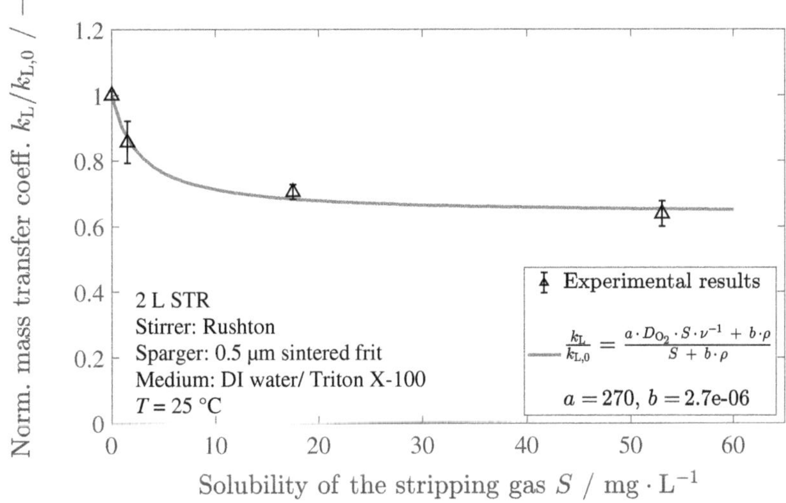

Fig. 5.33 Correlation for the mass transfer coefficient as a function of the solubility of the stripping gas

Chapter 6

Conclusion

Overcoming present limitations and improving biocatalytic and chemical processes, is driving current research in the field of process engineering. The demand for aeration techniques achieving high mass transfer rates is still increasing. One approach towards higher mass transfer rates is based on reducing the bubble size which enhances the volume specific interfacial area as well as prolonging the residence time of the gaseous phase.

The smallest bubbles observed offer the highest volume specific interfacial areas. They are characterized by diameters smaller than 1 μm. The existence of these ultrafine bubbles is controversially discussed in the scientific community, due to the fact that applying conventional thermodynamical and physical models lead to contradicting results. Using state of the art measurement techniques such as nanoparticle tracking analysis, first experimental results are presented, pointing towards their existence. The results show that with the current technical possibilities the highest achievable bubble density is limited, resulting in gas hold-ups smaller than $5 \cdot 10^{-5}\%$. Assuming an ongoing technical progress, gas hold-ups of 1% and higher can be technically achieved in the near future. Ultrafine bubbles can then contribute to optimize current processes. So far, at the current state of the art it is not beneficial to apply ultrafine bubbles in industrial processes.

In contrast to ultrafine bubbles, the aeration with fine bubbles at microscales is already an alternative to conventional aeration techniques. The impact of fine bubble aeration on hydrodynamics and mass transfer rates is experimentally investigated in this work considering three different laboratory scale reactor systems: bubble column, jet and stirred tank reactor. The stirred tank reactor shows the highest benefits through fine bubble aeration in regard to energy efficiency and achievable mass transfer rates.

Due to small initial bubble diameters, there is no significant influence of the power input on the bubble size distribution within the turbulent regime and a homogeneous distribution of the gaseous phase is present throughout the whole reactor. Thus, the stirred system can be operated at low power inputs, focusing on an efficient mixing within the tank. In addition, a decoupling of the usually linked hydrodynamics and mass transfer characteristics within the reactor is achieved. Compared to conventionally aerated stirred tank reactors, the complexity of the fine bubble aerated system is significantly lower. This reduces the effort for simulating and validating the reactor immensely, shortening development times and lowering costs for reactor design processes.

Regarding the mass transfer performance of the stirred tank reactor, fine bubble aeration leads to mass transfer coeffcients two times greater than those of conventionally aerated systems. Assuming a fixed mass transfer rate, measured mass transfer coefficients show, that fine bubble aeration better utilizes the gaseous phase. Compared to conventional aeration, up to 60% of the gas is saved, reaching the same $k_{\mathrm{L}}a$ values highlighting the potential of fine bubble aeration in terms of saving resources. Besides, the large volume specific interfacial surface area and the low rising velocity, it is found that the Laplace pressure enhances the mass transfer at microscales.

The importance of the Laplace pressure for the mass transfer rates of fine bubbles and its sensitivity to the bubble size is experimentally analyzed taking into account the impact of counterdiffusion effects. Further dissolved gases being present in the liquid reduce the concentration of the process gas within the bubble during the mass transfer process, decreasing the overall mass transfer rate. Applying established laser optical measurement techniques, the dissolution process of single microscale bubbles is observed. Using oxygen, as an often used reactant, the concentration gradients in the vicinity of the bubble are quantified. A model is developed describing the decreasing oxygen concentration within the bubble over time, while considering occurring counterdiffusion effects. The impact of counterdiffusion is further quantified for bubble swarms, investigating different bubble sizes and shapes. The influence of the degassing method on measured mass transfer coefficients is quantified, by comparing different commonly used physical degassing methods: boiling as well as gas stripping with argon, nitrogen or helium. In combination with the knowledge of the bubble size distribution during the mass transfer process, a novel model is introduced. It allows calculating mass transfer coefficients, considering the solubility of the stripping gas.

The insights gained in this work, enable a more precise prediction of mass transfer coefficients, no longer neglecting the impact of the Laplace pressure at microscales. The developed models should be implemented into process simulations, so in the future real systems can be considered capturing the influence of all substances on the mass transfer rates. In order to exploit the full potential of fine bubble aeration in the near furture, especially for biocatalytic reactions, investigations of local phenomena at the gas-liquid interface with respect to the influence of surface-active enzymes are necessary.

References

[Alh16] Alheshibri, M., Qian, J., Jehannin, M. and Craig, V.S. *A history of nanobubbles. Langmuir*, 32(43):11086–11100, 2016.

[Alv02] Alves, S., Maia, C., Vasconcelos, J. and Serralheiro, A. *Bubble size in aerated stirred tanks. Chemical Engineering Journal*, 89(1-3):109–117, 2002.

[AS19] Albert-Smet, I., Marcos-Vidal, A., Vaquero, J.J., Desco, M., Muñoz-Barrutia, A. and Ripoll, J. *Applications of light-sheet microscopy in microdevices. Frontiers in Neuroanatomy*, 13:1, 2019.

[Att13] Attard, P. *The stability of nanobubbles. The European Physical Journal Special Topics*, pp. 1–22, 2013.

[Bae94] Baehr, H.D. and Stephan, K. *Wärme-und stoffübertragung*, vol. 7. Springer, 1994.

[Bal12] Ball, P. *Nanobubbles are not a superficial matter. ChemPhysChem*, 13(8):2173–2177, 2012.

[Bat63] Bates, R.L., Fondy, P.L. and Corpstein, R.R. *Examination of some geometric parameters of impeller power. Industrial & Engineering Chemistry Process Design and Development*, 2(4):310–314, 1963.

[Bod17] Bodnár, T., Galdi, G.P. and Nečasová, Š. *Particles in Flows*. Springer, 2017.

[Bot16] Bothe, M. *Experimental Analysis and Modeling of Industrial Two-Phase Flows in Bubble Column Reactors*, vol. 1. Cuvillier Verlag, 2016.

[Bou01] Bouaifi, M., Hebrard, G., Bastoul, D. and Roustan, M. *A comparative study of gas hold-up, bubble size, interfacial area and mass transfer coefficients in stirred gas–liquid reactors and bubble columns. Chemical engineering and processing: Process intensification*, 40(2):97–111, 2001.

[Bra71] Brauer, H. *Grundlagen der Einphasen-und Mehrphasenströmungen*, vol. 2. Sauerländer, 1971.

[Bre98] Bredwell, M.D. and Worden, R.M. *Mass-transfer properties of microbubbles. 1. Experimental studies. Biotechnology progress*, 14(1):31–38, 1998.

[Bre08] Brenner, M.P. and Lohse, D. *Dynamic equilibrium mechanism for surface nanobubble stabilization. Physical review letters*, 101(21):214505, 2008.

[Buj87] Bujalski, W., Nienow, A., Chatwin, S. and Cooke, M. *The dependency on scale of power numbers of Rushton disc turbines. Chemical Engineering Science*, 42(2):317–326, 1987.

[Bun09] Bunkin, N., Suyazov, N., Shkirin, A., Ignatiev, P. and Indukaev, K. *Nanoscale structure of dissolved air bubbles in water as studied by measuring the elements of the scattering matrix. The Journal of chemical physics*, 130(13):134308, 2009.

[Bur97] Burns, S., Yiacoumi, S. and Tsouris, C. *Microbubble generation for environmental and industrial separations. Separation and Purification Technology*, 11(3):221–232, 1997.

[Cal98] Calliada, F., Campani, R., Bottinelli, O., Bozzini, A. and Sommaruga, M.G. *Ultrasound contrast agents: basic principles. European journal of radiology*, 27:S157–S160, 1998.

[Cho05] Cho, S.H., Kim, J.Y., Chun, J.H. and Kim, J.D. *Ultrasonic formation of nanobubbles and their zeta-potentials in aqueous electrolyte and surfactant solutions. Colloids and Surfaces A: Physicochemical and Engineering Aspects*, 269(1-3):28–34, 2005.

[Chr09] Christen, D.S. *Praxiswissen der chemischen verfahrenstechnik: handbuch für chemiker und verfahrensingenieure.* Springer-Verlag, 2009.

[Chu07] Chu, L.B., Xing, X.H., Yu, A.F., Zhou, Y.N., Sun, X.L. and Jurcik, B. *Enhanced ozonation of simulated dyestuff wastewater by microbubbles. Chemosphere*, 68(10):1854–1860, 2007.

[Chu08] Chu, L.B., Xing, X.H., Yu, A.F., Sun, X.L. and Jurcik, B. *Enhanced treatment of practical textile wastewater by microbubble ozonation. Process Safety and Environmental Protection*, 86(5):389–393, 2008.

[Cli78] Clift, R., Grace, J. and Weber, M. *Bubbles, drops, and particles.* Dover Publ., 1978.

[Dan51] Danckwerts, P. *Significance of liquid-film coefficients in gas absorption. Industrial & Engineering Chemistry*, 43(6):1460–1467, 1951.

[Dem12] Demming, S., Peterat, G., Llobera, A., Schmolke, H., Bruns, A., Kohlstedt, M., Al-Halhouli, A.a., Klages, C.P., Krull, R. and Büttgenbach, S. *Vertical microbubble column–A photonic lab-on-chip for cultivation and online analysis of yeast cell cultures. Biomicrofluidics*, 6(3):034106, 2012.

[DJ91] De Jong, N., Ten Cate, F., Lancee, C., Roelandt, J. and Bom, N. *Principles and recent developments in ultrasound contrast agents. Ultrasonics*, 29(4):324–330, 1991.

[Doi05] Doig, S.D., Diep, A. and Baganz, F. *Characterisation of a novel miniaturised bubble column bioreactor for high throughput cell cultivation. Biochemical engineering journal*, 23(2):97–105, 2005.

[Dor13] Doran, P.M. *Bioprocess engineering principles, second edition.* Elsevier, 2013.

[DR94] De Rijk, S.E., den Blanken, J.G. *et al.*. *Bubble size in flotation thickening*. *Water Research*, 28(2):465–473, 1994.

[Dru15] Druzinec, D., Salzig, D., Kraume, M. and Czermak, P. *Micro-bubble aeration in turbulent stirred bioreactors: Coalescence behavior in Pluronic F68 containing cell culture media*. *Chemical Engineering Science*, 126:160–168, 2015.

[Duc09] Ducker, W.A. *Contact angle and stability of interfacial nanobubbles*. *Langmuir*, 25(16):8907–8910, 2009.

[Duv12] Duval, E., Adichtchev, S., Sirotkin, S. and Mermet, A. *Long-lived submicrometric bubbles in very diluted alkali halide water solutions*. *Physical Chemistry Chemical Physics*, 14(12):4125–4132, 2012.

[Ebi13] Ebina, K., Shi, K., Hirao, M., Hashimoto, J., Kawato, Y., Kaneshiro, S., Morimoto, T., Koizumi, K. and Yoshikawa, H. *Oxygen and air nanobubble water solution promote the growth of plants, fishes, and mice*. *PLoS One*, 8(6):e65339, 2013.

[Ein56] Einstein, A. *Investigations on the Theory of the Brownian movement*. Courier Corporation, 1956.

[Eps50] Epstein, P.S. and Plesset, M.S. *On the stability of gas bubbles in liquid-gas solutions*. *The Journal of Chemical Physics*, 18(11):1505–1509, 1950.

[Fil10] Filipe, V., Hawe, A. and Jiskoot, W. *Critical evaluation of Nanoparticle Tracking Analysis (NTA) by NanoSight for the measurement of nanoparticles and protein aggregates*. *Pharmaceutical research*, 27(5):796–810, 2010.

[Fra00] Fradin, C., Braslau, A., Luzet, D., Smilgies, D., Alba, M., Boudet, N., Mecke, K. and Daillant, J. *Reduction in the surface energy of liquid interfaces at short length scales*. *Nature*, 403(6772):871–874, 2000.

[Gei91] Geisler, R., Krebs, R. and Forschner, P. *Local Turbulent Shear Stress in Stirred Vessels and its Significance for Different Mixing Tasks*. In *Proc of 8th European Conf on Mixing*, pp. 241–251. Inst of Chem Eng, Rugby, UK, 1991.

[Haa10] Haapala, A., Honkanen, M., Liimatainen, H., Stoor, T. and Niinimäki, J. *Hydrodynamic drag and rise velocity of microbubbles in papermaking process waters*. *Chemical Engineering Journal*, 162(3):956–964, 2010.

[Had11] Hadamard, J.S. *Slow Permanent Movement of a Liquid and Viscous Sphere in a Viscous Liquid*. *CR Hebd. Seances Acad. Sci. Paris*, 152:1735–1738, 1911.

[Han17] Han, Y., Liu, Y., Jiang, H., Xing, W. and Chen, R. *Large scale preparation of microbubbles by multi-channel ceramic membranes: hydrodynamics and mass transfer characteristics*. *The Canadian Journal of Chemical Engineering*, 95(11):2176–2185, 2017.

[He05] He, S. and Attard, P. *Surface tension of a Lennard-Jones liquid under supersaturation*. *Physical Chemistry Chemical Physics*, 7(15):2928–2935, 2005.

[Hig35] Higbie, R. *The rate of absorption of a pure gas into a still liquid during short periods of exposure. Trans. AIChE*, 31:365–389, 1935.

[Hol13] Hole, P., Sillence, K., Hannell, C., Maguire, C.M., Roesslein, M., Suarez, G., Capracotta, S., Magdolenova, Z., Horev-Azaria, L., Dybowska, A. *et al.*. *Interlaboratory comparison of size measurements on nanoparticles using nanoparticle tracking analysis (NTA). Journal of Nanoparticle Research*, 15(12):2101, 2013.

[Hon84] Hong, W.H. and Brauer, H. *Stoffaustausch zwischen Gas und Flüssigkeit in Blasensäulen: mit 5 Tabellen.* VDI-Verlag, 1984.

[Ish00] Ishida, N., Inoue, T., Miyahara, M. and Higashitani, K. *Nano bubbles on a hydrophobic surface in water observed by tapping-mode atomic force microscopy. Langmuir*, 16(16):6377–6380, 2000.

[Iwa17] Iwakiri, M., Terasaka, K., Fujioka, S., Schluter, M., Kastens, S. and Tanaka, S. *Mass Transfer from a Shrinking Single Microbubble Rising in Water. Japanese Journal of Multiphase Flow*, 30(5):529–535, 2017.

[Jin07] Jin, F., Ye, X. and Wu, C. *Observation of kinetic and structural scalings during slow coalescence of nanobubbles in an aqueous solution. The Journal of Physical Chemistry B*, 111(46):13143–13146, 2007.

[Jon99] Jones, S., Evans, G. and Galvin, K. *Bubble nucleation from gas cavities—a review. Advances in colloid and interface science*, 80(1):27–50, 1999.

[Jud76] Judat, H. *Eignung schnellaufender Ruehrertypen zum Dispergieren von Gasen in niedrig-viskosen Fluessigkeiten. Chemie Ingenieur Technik*, 48:228–229, 1976.

[Kar97] Karasso, P. and Mungal, M. *PLIF measurements in aqueous flows using the Nd: YAG laser. Experiments in fluids*, 23(5):382–387, 1997.

[Kay86] Kayser, R. *Effect of capillary waves on surface tension. Physical Review A*, 33(3):1948, 1986.

[Kel96] Kelsall, G.H., Tang, S., Yurdakul, S. and Smith, A.L. *Electrophoretic behaviour of bubbles in aqueous electrolytes. Journal of the Chemical Society, Faraday Transactions*, 92(20):3887–3893, 1996.

[Kik09] Kikuchi, K., Ioka, A., Oku, T., Tanaka, Y., Saihara, Y. and Ogumi, Z. *Concentration determination of oxygen nanobubbles in electrolyzed water. Journal of colloid and interface science*, 329(2):306–309, 2009.

[Kim00] Kim, J.Y., Song, M.G. and Kim, J.D. *Zeta potential of nanobubbles generated by ultrasonication in aqueous alkyl polyglycoside solutions. Journal of colloid and interface science*, 223(2):285–291, 2000.

[Kra12] Kraume, M. *Transportvorgänge in der Verfahrenstechnik Grundlagen und apparative Umsetzungen, 2. bearb. Auflage.* Springer-Vieweg Wiesbaden, 2012.

[Kru20] Krull, R., Hempel, D. and Wucherpfennig, T. *Bioverfahrenstechnik.* In *Dubbel Taschenbuch für den Maschinenbau 3: Maschinen und Systeme*, pp. 565–600. Springer, 2020.

[Kuk06] Kukizaki, M. *Effect of wettability of porous glass membrane to water phase on monodispersed nanobubble formation. Progress in Multiphase Flow Research*, 1:33–41, 2006.

[Kuk09] Kukizaki, M. *Microbubble formation using asymmetric Shirasu porous glass (SPG) membranes and porous ceramic membranes—A comparative study. Colloids and Surfaces A: Physicochemical and Engineering Aspects*, 340(1-3):20–32, 2009.

[Lei94] Leighton, T. *Acoustic bubble detection-I. The detection of stable gas bodies. Environmental engineering*, 7:9–16, 1994.

[Lew24] Lewis, W. and Whitman, W. *Principles of gas absorption. Industrial & Engineering Chemistry*, 16(12):1215–1220, 1924.

[Li18] Li, G., Li, H., Wei, G., He, X., Xu, S., Chen, K., Ouyang, P. and Ji, X. *Hydrodynamics, mass transfer and cell growth characteristics in a novel microbubble stirred bioreactor employing sintered porous metal plate impeller as gas sparger. Chemical Engineering Science*, 192:665–677, 2018.

[Liu13a] Liu, C., Tanaka, H., Zhang, J., Zhang, L., Yang, J., Huang, X. and Kubota, N. *Successful application of Shirasu porous glass (SPG) membrane system for microbubble aeration in a biofilm reactor treating synthetic wastewater. Separation and Purification Technology*, 103:53–59, 2013.

[Liu13b] Liu, S., Kawagoe, Y., Makino, Y. and Oshita, S. *Effects of nanobubbles on the physicochemical properties of water: The basis for peculiar properties of water containing nanobubbles. Chemical Engineering Science*, 93:250–256, 2013.

[Lju97] Ljunggren, S. and Eriksson, J.C. *The lifetime of a colloid-sized gas bubble in water and the cause of the hydrophobic attraction. Colloids and Surfaces A: Physicochemical and Engineering Aspects*, 129:151–155, 1997.

[Loh15] Lohse, D., Zhang, X. *et al.*. *Pinning and gas oversaturation imply stable single surface nanobubbles. Physical Review E*, 91(3):031003, 2015.

[Lou00] Lou, S.T., Ouyang, Z.Q., Zhang, Y., Li, X.J., Hu, J., Li, M.Q. and Yang, F.J. *Nanobubbles on solid surface imaged by atomic force microscopy. Journal of Vacuum Science & Technology B: Microelectronics and Nanometer Structures Processing, Measurement, and Phenomena*, 18(5):2573–2575, 2000.

[Mae15] Maeda, Y., Hosokawa, S., Baba, Y., Tomiyama, A. and Ito, Y. *Generation mechanism of micro-bubbles in a pressurized dissolution method. Experimental Thermal and Fluid Science*, 60:201–207, 2015.

[Mah15] Mahmood, K.A., Wilkinson, S.J. and Zimmerman, W.B. *Airlift bioreactor for biological applications with microbubble mediated transport processes. Chemical Engineering Science*, 137:243–253, 2015.

[Mat18] Matthes, S., Kastens, S., Thomas, B., Ohde, D., Bubenheim, P., Liese, A., Tanaka, S., Terasaka, K. and Schlüter, M. *Characterization of Fine Bubbles for Biocatalytic Processes.* In *Proc. of 19th Int. Symp. on Appl. of Laser and Imaging Tech. to Fluid Mech.*, pp. 1299–1312. Lisbon, Portugal, ISBN: 978-989-20-9177-8, 2018.

[Mat20] Matthes, S., Thomas, B., Ohde, D., Hoffmann, M., Bubenheim, P., Liese, A., Tanaka, S., Terasaka, K. and Schlueter, M. *Hydrodynamic and Mass Transfer Correlation in a Microbubble Aerated Stirred Tank Reactor. Journal of Chemical Engineering of Japan*, 53(10):577–584, 2020.

[Mic18] Michelin, S., Guérin, E. and Lauga, E. *Collective dissolution of microbubbles. Physical Review Fluids*, 3(4):043601, 2018.

[Mil81] Miller, D. *Ultrasonic detection of resonant cavitation bubbles in a flow tube by their second-harmonic emissions. Ultrasonics*, 19(5):217–224, 1981.

[Moo03] Moody, M.P. and Attard, P. *Curvature-dependent surface tension of a growing droplet. Physical review letters*, 91(5):056104, 2003.

[Moo04] Moody, M.P. and Attard, P. *Monte Carlo simulation methodology of the ghost interface theory for the planar surface tension. The Journal of chemical physics*, 120(4):1892–1904, 2004.

[Mur13] Murphy, D.B. and Davidson, M.W. *Fundamentals of Light Microscopy and Electronic Imaging.* John Wiley & Sons, 2013.

[Nag92] Nagahiro, J., Iwamoto, J. and Higuchi, K. *Experiments for fine air bubble production in liquids using ejectors. Transactions of the ASAE*, 35(5):1581–1590, 1992.

[Nak13] Nakatake, Y., Kisu, S., Shigyo, K., Eguchi, T. and Watanabe, T. *Effect of nano air-bubbles mixed into gas oil on common-rail diesel engine. Energy*, 59:233–239, 2013.

[Neh04] Nehring, D., Czermak, P., Vorlop, J. and Lübben, H. *Experimental study of a ceramic microsparging aeration system in a pilot-scale animal cell culture. Biotechnology progress*, 20(6):1710–1717, 2004.

[Oh13] Oh, S.H., Yoon, S.H., Song, H., Han, J.G. and Kim, J.M. *Effect of hydrogen nanobubble addition on combustion characteristics of gasoline engine. International journal of hydrogen energy*, 38(34):14849–14853, 2013.

[Oh17] Oh, S.H. and Kim, J.M. *Generation and stability of bulk nanobubbles. Langmuir*, 33(15):3818–3823, 2017.

[Ohd19] Ohde, D., Thomas, B., Matthes, S., Percin, Z., Engelmann, C., Bubenheim, P., Terasaka, K., Schlüter, M. and Liese, A. *Fine Bubble-based CO2 Capture Mediated by Triethanolamine Coupled to Whole Cell Biotransformation. Chemie Ingenieur Technik*, 91(12):1822–1826, 2019.

[Ohg10] Ohgaki, K., Khanh, N.Q., Jodon, Y., Touji, A. and Nakagawa, T. *Physicochemical approach to nanobubble solutions. Chemical Engineering Science*, 65(3):1296–1300, 2010.

[Ola18] Olarte, O.E., Andilla, J., Gualda, E.J. and Loza-Alvarez, P. *Light-sheet microscopy: a tutorial. Advances in Optics and Photonics*, 10(1):111–179, 2018.

[Ona99] Onari, H., Saga, T., Watanabe, K., Maeda, K. and Matsuo, K. *High functional characteristics of micro-bubbles and water purification. Resources processing*, 46(4):238–244, 1999.

[Par94a] Parker, J.L., Claesson, P.M. and Attard, P. *Bubbles, cavities, and the long-ranged attraction between hydrophobic surfaces. The Journal of Physical Chemistry*, 98(34):8468–8480, 1994.

[Par94b] Parthasarathy, R. and Ahmed, N. *Bubble size distribution in a gas sparged vessel agitated by a Rushton turbine. Industrial & engineering chemistry research*, 33(3):703–711, 1994.

[Par08] Parkinson, L., Sedev, R., Fornasiero, D. and Ralston, J. *The terminal rise velocity of 10–100 μm diameter bubbles in water. Journal of colloid and interface science*, 322(1):168–172, 2008.

[Par11] Parkinson, L. and Ralston, J. *Dynamic aspects of small bubble and hydrophilic solid encounters. Advances in Colloid and Interface Science*, 168(1-2):198–209, 2011.

[Pee53] Peebles, F. and Garber, H. *Studies on the motion of gas bubbles in liquid. Chem. Eng. Prog. Symp.*, 49(2):88–97, 1953.

[Pet14] Peterat, G., Schmolke, H., Lorenz, T., Llobera, A., Rasch, D., Al-Halhouli, A.T., Dietzel, A., Büttgenbach, S., Klages, C.P. and Krull, R. *Characterization of oxygen transfer in vertical microbubble columns for aerobic biotechnological processes. Biotechnology and bioengineering*, 111(9):1809–1819, 2014.

[Pop00] Pope, S.B. *Turbulent flows.* Camebridge University Press, Camebridge, U.K., 2000.

[Qi02] Qi, H. *Membrane and micro-sparging aerations in long-term high-density perfusion cultures of animal cells.* 2002.

[Räb13] Räbiger, N. and Schlüter, M. *Bildung und Bewegung von Tropfen und Blasen. VDI Wärmeatlas*, 2013.

[Ric54] Richardson, J. and Zaki, W. *Part1, sedimentation and fluidisation. Ind. Eng. Chem.*, 41:S35–S53, 1954.

[Ros19] Rosseburg, A. *Influence of heterogeneous bubbly flows on mixing and mass transfer performance in stirred tanks for mammalian cell cultivation: A study in transparent 3 L and 12 000 L reactors*, vol. 5. Cuvillier Verlag, 2019.

[Rüt18] Rüttinger, S., Spille, C., Hoffmann, M. and Schlüter, M. *Laser-induced fluorescence in multiphase systems. ChemBioEng Reviews*, 5(4):253–269, 2018.

[Ryb11] Rybczynski, W. *On the Translatory Motion of a Fluid Sphere in a Viscous Medium. Bull. Acad. Sci. Cracow (A)*, 40:33–78, 1911.

[Sam96] Sam, A., Gomez, C. and Finch, J. *Axial velocity profiles of single bubbles in water/frother solutions. International Journal of Mineral Processing*, 47(3-4):177–196, 1996.

[Sas14] Sasic, S., Sibaki, E.K. and Ström, H. *Direct numerical simulation of a hydrodynamic interaction between settling particles and rising microbubbles. European Journal of Mechanics-B/Fluids*, 43:65–75, 2014.

[Sat12] Sattler, K. *Thermische Trennverfahren: Grundlagen, Auslegung, Apparate.* John Wiley & Sons, 2012.

[Sch95] Schneider, M., Arditi, M., Barrau, M.B., Brochot, J., Broillet, A., Ventrone, R. and Yan, F. *BR1: a new ultrasonographic contrast agent based on sulfur hexafluoride-filled microbubbles. Investigative radiology*, 30(8):451–457, 1995.

[Sed11] Seddon, J.R., Zandvliet, H.J. and Lohse, D. *Knudsen gas provides nanobubble stability. Physical review letters*, 107(11):116101, 2011.

[Sed12] Seddon, J.R., Lohse, D., Ducker, W.A. and Craig, V.S. *A deliberation on nanobubbles at surfaces and in bulk. ChemPhysChem*, 13(8):2179–2187, 2012.

[Shi97] Shi, W.T., Forsberg, F. and Oung, H. *Spectral broadening in conventional and harmonic Doppler measurements with gaseous contrast agents.* In *1997 IEEE Ultrasonics Symposium Proceedings. An International Symposium (Cat. No. 97CH36118)*, vol. 2, pp. 1575–1578. IEEE, 1997.

[Shi15] Shin, D., Park, J.B., Kim, Y.J., Kim, S.J., Kang, J.H., Lee, B., Cho, S.P., Hong, B.H. and Novoselov, K.S. *Growth dynamics and gas transport mechanism of nanobubbles in graphene liquid cells. Nature communications*, 6(1):1–6, 2015.

[Sto51] Stokes, G.G. *On the Effect of the Internal Friction of Fluids on the Motion of Pendulums.* Pitt Press Cambridge, 1851.

[Str77] Streicher, R. and Schügerl, K. *Interchange of fluid mechanics and mass exchange in droplets. Chemical Engineering Science*, 32(1):23–33, 1977.

[Sus89] Suslick, K.S. *The chemical effects of ultrasound. Scientific American*, 260(2):80–87, 1989.

[Ter11] Terasaka, K., Hirabayashi, A., Nishino, T., Fujioka, S. and Kobayashi, D. *Development of microbubble aerator for waste water treatment using aerobic activated sludge. Chemical engineering science*, 66(14):3172–3179, 2011.

[Tho20] Thomas, B., Ohde, D., Matthes, S., Engelmann, C., Bubenheim, P., Terasaka, K., Schlüter, M. and Liese, A. *Comparative investigation of fine bubble and macrobubble aeration on gas utility and biotransformation productivity. Biotechnology and bioengineering*, 2020.

[Tim18] Timmermann, J. *Experimental analysis of fast reactions in gas liquid flows*, vol. 3. Cuvillier Verlag, 2018.

[Tol49] Tolman, R.C. *The superficial density of matter at a liquid-vapor boundary. The journal of chemical physics*, 17(2):118–127, 1949.

[Tsu14] Tsuge, H. *Micro-and Nanobubbles: Fundamentals and Applications.* CRC press, 2014.

[Uch11] Uchida, T., Oshita, S., Ohmori, M., Tsuno, T., Soejima, K., Shinozaki, S., Take, Y. and Mitsuda, K. *Transmission electron microscopic observations of nanobubbles and their capture of impurities in wastewater. Nanoscale research letters*, 6(1):295, 2011.

[Ush10] Ushikubo, F.Y., Furukawa, T., Nakagawa, R., Enari, M., Makino, Y., Kawagoe, Y., Shiina, T. and Oshita, S. *Evidence of the existence and the stability of nanobubbles in water. Colloids and Surfaces A: Physicochemical and Engineering Aspects*, 361(1-3):31–37, 2010.

[Ush11] Ushida, A., Hasegawa, T., Amaki, K., Nakajima, T., Takahashi, N. and Narumi, T. *Investigation on Washing Effects for Nano-Bubble/Surfactant Mixtures in an Alternating Flow. Trans. Jpn. Soc. Mech. Eng. B*, 77:1219–1228, 2011.

[Ush12] Ushida, A., Hasegawa, T., Nakajima, T., Uchiyama, H. and Narumi, T. *Drag reduction effect of nanobubble mixture flows through micro-orifices and capillaries. Experimental thermal and fluid science*, 39:54–59, 2012.

[Ush13] Ushida, A., Hasegawa, T., Narumi, T. and Nakajima, T. *Flow properties of nanobubble mixtures passing through micro-orifices. International journal of heat and fluid flow*, 40:106–115, 2013.

[vL11] van Limbeek, M.A. and Seddon, J.R. *Surface nanobubbles as a function of gas type. Langmuir*, 27(14):8694–8699, 2011.

[Voo85] Voorhees, P.W. *The theory of Ostwald ripening. Journal of Statistical Physics*, 38(1-2):231–252, 1985.

[Wan13] Wang, D. and Fan, L.S. *Particle characterization and behavior relevant to fluidized bed combustion and gasification systems.* In *Fluidized bed technologies for near-zero emission combustion and gasification*, pp. 42–76. Elsevier, 2013.

[Wan19] Wang, Q., Zhao, H., Qi, N., Qin, Y., Zhang, X. and Li, Y. *Generation and stability of size-adjustable bulk nanobubbles based on periodic pressure change. Scientific reports*, 9(1):1–9, 2019.

[Was87] Wasowski, T. and Blaß, E. *Wake-Phänomene hinter festen und fluiden Partikeln. Chemie Ingenieur Technik*, 59(7):544–555, 1987.

[Wei12] Weijs, J.H., Seddon, J.R. and Lohse, D. *Diffusive shielding stabilizes bulk nanobubble clusters. ChemPhysChem*, 13(8):2197–2204, 2012.

[Wei13] Weijs, J.H. and Lohse, D. *Why surface nanobubbles live for hours. Physical review letters*, 110(5):054501, 2013.

[Wor98] Worden, R.M. and Bredwell, M.D. *Mass-Transfer Properties of Microbubbles. 2. Analysis Using a Dynamic Model. Biotechnology Progress*, 14(1):39–46, 1998.

[Wu08] Wu, Z., Chen, H., Dong, Y., Mao, H., Sun, J., Chen, S., Craig, V.S. and Hu, J. *Cleaning using nanobubbles: defouling by electrochemical generation of bubbles. Journal of colloid and interface science*, 328(1):10–14, 2008.

[Yas16] Yasui, K., Tuziuti, T., Kanematsu, W. and Kato, K. *Dynamic equilibrium model for a bulk nanobubble and a microbubble partly covered with hydrophobic material. Langmuir*, 32(43):11101–11110, 2016.

[Yas19] Yasuda, K., Matsushima, H. and Asakura, Y. *Generation and reduction of bulk nanobubbles by ultrasonic irradiation. Chemical Engineering Science*, 195:455–461, 2019.

[Yin13] Ying, K., Al-Mashhadani, M.K., Hanotu, J.O., Gilmour, D.J. and Zimmerman, W.B. *Enhanced mass transfer in microbubble driven airlift bioreactor for microalgal culture. Engineering*, 5(9):735–743, 2013.

[Zha06a] Zhang, X.H., Li, G., Maeda, N. and Hu, J. *Removal of induced nanobubbles from water/graphite interfaces by partial degassing. Langmuir*, 22(22):9238–9243, 2006.

[Zha06b] Zhang, X.H., Maeda, N. and Craig, V.S. *Physical properties of nanobubbles on hydrophobic surfaces in water and aqueous solutions. Langmuir*, 22(11):5025–5035, 2006.

[Zha07] Zhang, X.H., Khan, A. and Ducker, W.A. *A nanoscale gas state. Physical review letters*, 98(13):136101, 2007.

[Zim09] Zimmerman, W.B., Hewakandamby, B.N., Tesař, V., Bandulasena, H.H. and Omotowa, O.A. *On the design and simulation of an airlift loop bioreactor with microbubble generation by fluidic oscillation. Food and Bioproducts Processing*, 87(3):215–227, 2009.

[Zlo67] Zlokarnik, M. and Judat, H. *Rohr-und Scheibenrührer—zwei leistungsfähige Rührer zur Flüssigkeitsbegasung. Chemie Ingenieur Technik*, 39(20):1163–1168, 1967.

[Zlo01] Zlokarnik, M. *Stirring, general. Stirring. Weinheim, Germany: Wiley-VCH Verlag GmbH*, pp. 1–75, 2001.

Appendix A

Technical data of the SOPAT probes

Table A.1 Technical data of the SOPAT probes Pl and Sc used for fine bubble characterization

Product Category	SOPAT-VI	
Product Model	**Pl**	**Sc**
Measurement Range / µm	3 - 350	9 - 1 200
Field of View / mm	0.8	2.7
Tube Length / mm	500	
Tube Diameter / mm	12	
Pressure Range / bar	-0.1 - 320	
Probe Temperature Range / °C	-50 - 450	
Periphery Temperature Range / °C	0 - 50	
pH-Level	0 - 14	
Probe Window Material	Sapphire Glass	
Probe Tube Material	Hastelloy C22	
Probe Housing Material	1.4404 (316 L)	
Weight (without Cable) / kg	4	
Focus	Electronic	
Picture Rate / Hz	15	
Picture Resolution / MP	5	
Power Input / VA	140	
Certifications	CE, IP68, CIP/SIP, ATEX	

Appendix B

Additional data for the fine bubble aerated jet reactor setup

The figures B.1 and B.2 provide the logarithmic normal bubble size distributions for the swirl and the ejector nozzle fine bubble generator. The BSD are given for all analyzed volumetric power inputs $P \cdot V^{-1} \in \{6, 8, 10\}$ kW/m^3 and a fixed gassing rate of $\dot{V}_G = 100$ mL/min.

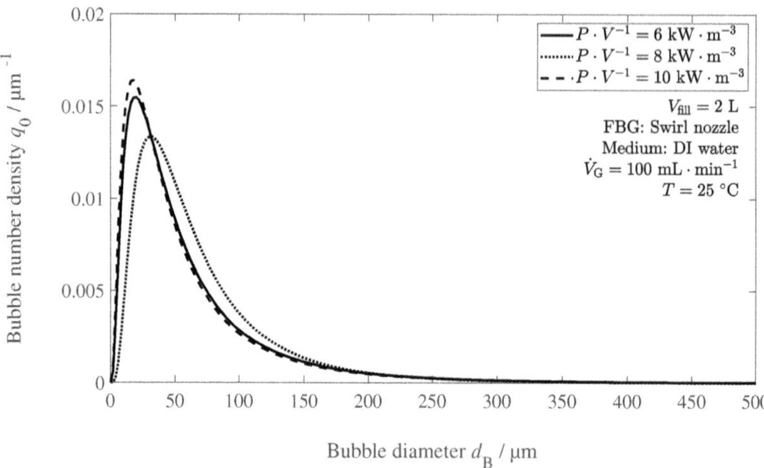

Fig. B.1 Logarithmic normal BSD for the swirl nozzle as a function of the power input $P \cdot V^{-1}$ for a fixed gassing rate of $\dot{V}_G = 100$ mL/min

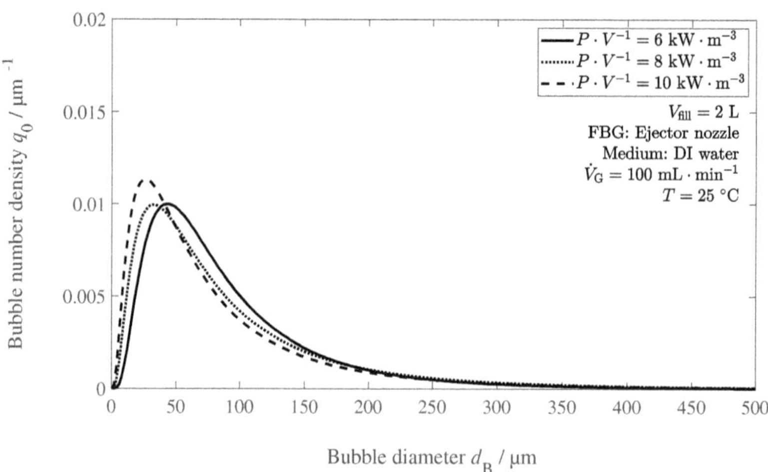

Fig. B.2 Logarithmic normal BSD for the ejector nozzle as a function of the power input $P \cdot V^{-1}$ for a fixed gassing rate of $\dot{V}_G = 100$ mL/min

Appendix C

Additional data for the membrane aerated STR

The figures C.1, C.2 and C.3 display the influence of the impeller geometry on the BSD in the membrane aerated STR. Comparing a Rushton turbine, a pitched blade impeller and a segment impeller, the BSD are measured at four different locations inside the vessel (compare figure 5.16) and for three different volumetric power inputs $P \cdot V^{-1} \in \{190, 640, 1525\}$ W/m^3.

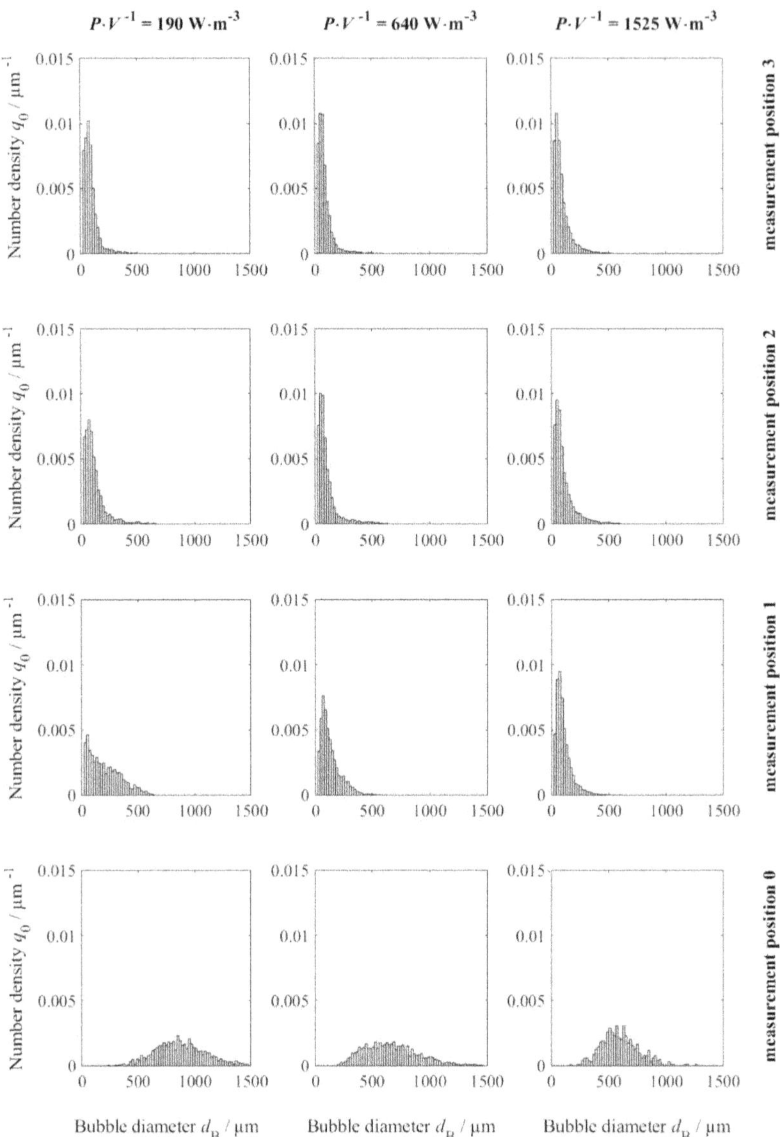

Fig. C.1 BSD in membrane aerated STR as a function of power input and measurement location for a single Rushton stirrer. Medium: DI water/Triton X-100, $\dot{V}_G = 100$ mL/min

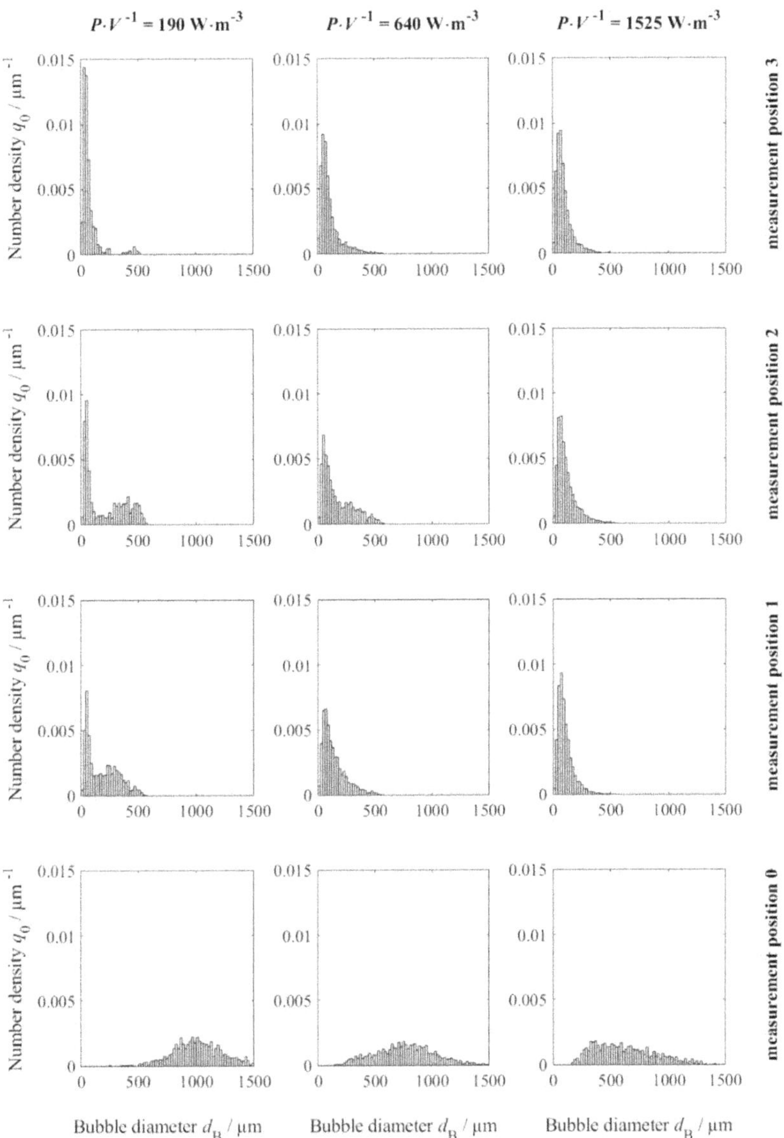

Fig. C.2 BSD in membrane aerated STR as a function of power input and measurement location for a single pitched blade stirrer. Medium: DI water/Triton X-100, $\dot{V}_G = 100$ mL/min

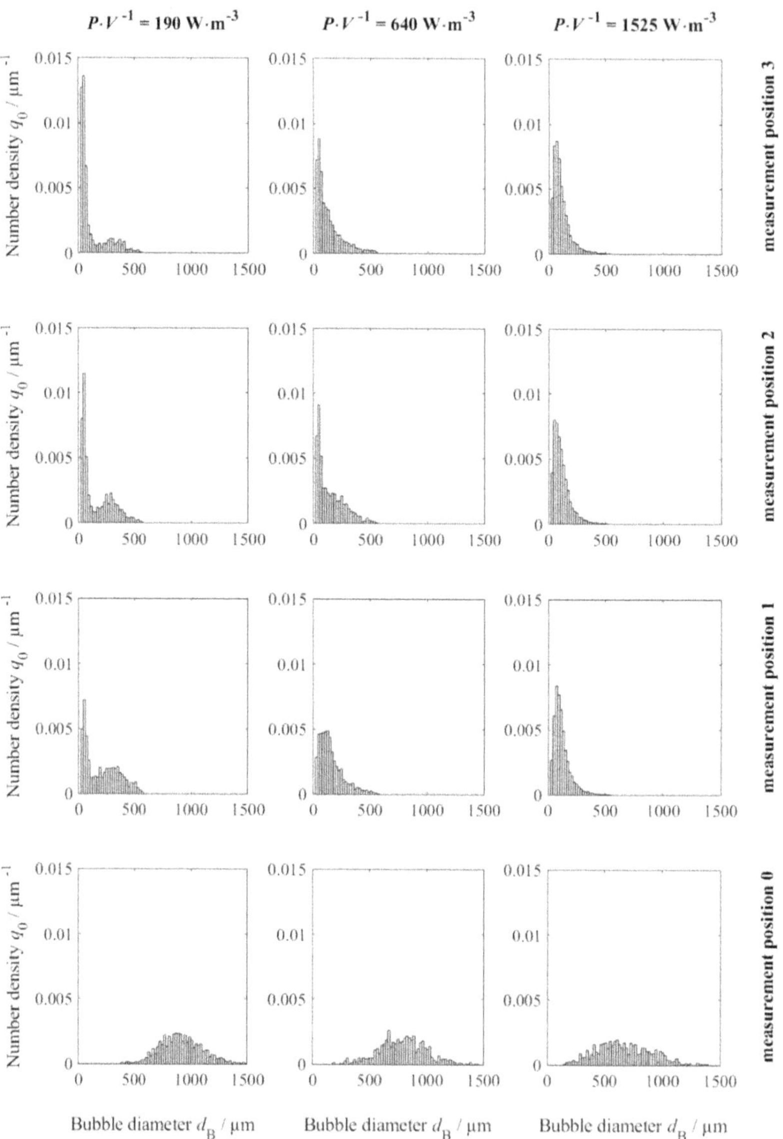

Fig. C.3 BSD in membrane aerated STR as a function of power input and measurement location for a single segment stirrer. Medium: DI water/Triton X-100, $\dot{V}_G = 100$ mL/min

www.ingramcontent.com/pod-product-compliance
Ingram Content Group UK Ltd.
Pitfield, Milton Keynes, MK11 3LW, UK
UKHW061658200726
13853UKWH00013B/2467